HISTORICIZING HUMANS

SCIENCE AND CULTURE IN THE NINETEENTH CENTURY

Bernard Lightman, Editor

HISTORICIZING HUMANS

Deep Time, Evolution, and Race in Nineteenth-Century British Sciences

EDITED BY EFRAM SERA-SHRIAR

WITH AN AFTERWORD BY THEODORE KODITSCHEK

University of Pittsburgh Press

Published by the University of Pittsburgh Press, Pittsburgh, Pa., 15260

Manufactured in the United States of America
Printed on acid-free paper
10 9 8 7 6 5 4 3 2 1

Cataloging-in-Publication data is available from the Library of Congress

ISBN 13: 978-0-8229-4529-1

Cover art: "Mammoth Hunt." By Ernest Griset (1860s/1870s).
Courtesy of Bromley Historic Collections and Professor Tim Murray.
Cover design by Joel W. Coggins

CONTENTS

ACKNOWLEDGMENTS

This volume builds on and moves beyond an earlier collection of papers that was published as a special issue in the journal *Studies in History and Philosophy of Biological and Biomedical Sciences* in 2015. My thanks extend to all of the contributors for their wonderful chapters. I am also grateful to Peter Kjaergaard and Jamie Elwick for their encouragement during the process of putting together this volume. Many thanks to Bernie Lightman, Fern Elsdon-Baker, and the rest of the team on the "Science and Religion: Exploring the Spectrum" project for their support in the early stages of my research on E. B. Tylor. Earlier versions of some of the chapters from this volume were presented at the annual conference of the British Society for the History of Science in Swansea, UK, in July 2015, and I am thankful to Adam Mosley for assisting in the organization of our panel. Finally, I am indebted to Abby Collier and the anonymous referees for helping to bring this volume to press.

HISTORICIZING HUMANS

INTRODUCTION

FROM THE BEGINNING

Human History Theories in Nineteenth-Century British Sciences

EFRAM SERA-SHRIAR

On the evening of 19 March 1858, the historian Henry Thomas Buckle (1821–1862) delivered a lecture at the Royal Institution (RI) on the topic of human history. Never before had Buckle delivered a public lecture, and there was so much interest in seeing his maiden disquisition that the doors of the RI were opened earlier than normal to accommodate a larger audience. In fact, demand for tickets was so high that even Buckle could not get enough tickets for his friends. Buckle's biographer, Alfred Henry Huth (1850–1910), stated that "the theatre was crammed from floor to ceiling by a brilliant and excited audience," which included some of the most influential gentlemen of science of the day such as Michael Faraday (1791–1867), Richard Owen (1804–1892), and Roderick Murchison (1792–1871). As he took to the stage, "the loud buzz of conversation was drowned in a burst of applause," and Buckle went on to give a hugely popular address.[1]

Along with his two-volume work, *History of Civilization in England* (1857), this lecture at the RI helped to springboard Buckle into a celebrity status.[2] The spectacle surrounding Buckle's lecture at the RI serves as a good example of how nineteenth-century British science was engrossed in studies of deep time and human history.[3] This captivation with humanity's past was not just something that scientific and medical researchers were interested in; a larger popular audience had an insatiable appetite for the topic as well. As Peter Bowler has argued, the nineteenth century was "an age dominated by a fascination with the past. History offered the preferred way of understanding how both human society and the material world operated."[4]

The aim in this volume is to look at some of the ways in which nineteenth-century scientific and medical researchers historicized humans within Britain and its empire. The historicization of humans within the context of this collection means the process of constructing human histories for various scientific, religious, and sociopolitical purposes. When it came to historicizing humans, nineteenth-century scientific and medical practitioners were varied in their methodological and theoretical approaches, and they utilized data from all over the world. While these varied approaches indicate that there was no absolute consensus on humanity's past, there were some underlying questions and shared assumptions in all nineteenth-century investigations into human history. Even competing theories such as monogenesis (the single origin of humans) and polygenesis (the multiple origins of humans) intersected in fascinating ways, with new commitments to contingency and chance and older ones such as providence and progress. Some of the key questions to emerge were as follows: Did the various races living throughout the world develop from a single location, or were their physical and social differences evidence for their separate genesis? Was it even possible to trace the development of humans or had too much time passed since the dawn of their emergence? How did new types of evolutionary theories transform nineteenth-century understandings of human ancestry? In this volume the aim is to examine these core questions about human history through an imperial, multidisciplinary perspective.

The word "historicize" emerged during the nineteenth century. The political theorist and classical historian George Grote (1794–1871) was the first Briton to use the term, and it appeared in his twelve-volume work, *A History of Greece* (1846–1856).[5] His usage, however, was indebted to earlier Italian, French, and German historical writing by renowned figures such as Giambattista Vico (1668–1744), Michel de Montaigne (1533–1592), Georg Wilhelm Friedrich Hegel (1770–1831), and Karl Wilhelm Friedrich Schlegel (1772–1829).[6] Nevertheless, the practice of historicizing humans took on numerous forms during the nineteenth century, and there were many different ways in which scientific and medical practitioners throughout the British Empire attempted to understand human origins and construct racial histories. Some of these theories upheld the older narratives based on biblical scripture, while others challenged these models, claiming that they were overly simplistic explanations of the Earth's history and ignored newly discovered evidence. The examination of human history, therefore, provides an important historiographical focal point, because both science and religion were interested in the origins of life and often offered competing explanations of human development. Thus there were many instances where the lines between science and religion were blurred.[7] This alleged tension between science and religion will be explored in the chapters that follow.

At a theoretical level Mark Bevir has argued that there were three broad underlying forms of historicism in the writings of most nineteenth-century figures: Whig historiography, which argued for an inevitable progression toward greater liberty and enlightenment; Romanticism, which glorified the past; and Positivism, which articulated a system that saw European civilizations progress from a theological stage to a metaphysical stage, before finally arriving at a positive or scientific stage. I would add evolutionism to this list, though, as Darwin's concept of natural selection, with its emphasis on random variation, is a form of historicism that is distinctly different from the other three versions.[8] At an evidentiary level, Martin Rudwick has argued that nineteenth-century practitioners interested in human history drew their data from three primary types of sources: texts (i.e., historical records, scripture, travelogues, etc.), human anatomy (i.e., skeletal remains, anthropometric measurements, etc.), and archaeological artifacts (i.e., stone tools, ruins, totems, etc.).[9] In the chapters that follow, we can see the application of all of these theories of historicism, and evidentiary sources, in the works of nineteenth-century scientific and medical researchers. Furthermore, debates over which types of theories and data sets were to be given priority when tracing human history were widespread in the nineteenth century. Examples of these discussions will feature throughout the volume.

Chronological periods also differed in length depending on the nature of an investigation. Many practitioners struggled to comprehend the implications of vastly expanded time frames. As John McPhee famously wrote, "Numbers do not seem to work well with regard to deep time. Any number above a couple of thousand years—fifty thousand, fifty million—will with nearly equal effect awe the imagination to the point of paralysis."[10] Yet not every study in the nineteenth century looked at the deep histories of human groups, and there were many instances where researchers historicized people over only short periods of time—such as a few centuries.[11] By contrast, there were also some researchers who challenged human developmental and evolutionary theories altogether. Figures such as the ethnologist and anatomist Robert Knox (1791–1862), believed that too much time had passed since the dawn of humans. He argued that only examinations of the current state of races could be substantiated empirically.[12] In his infamous book *Races of Man* (1850), Knox stated, "How worthless are these chronologies! How replete with error human history has been proved to be."[13] Regardless of their perspectives, though, the implication of all these investigations fed into larger discussions on human ancestry.

One of the core issues that was repeatedly discussed, challenged, renegotiated, and deconstructed, depending on new constellations of evidence, was special creation. During the nineteenth century, Britons witnessed a shift away from traditional biblical ideas about a separate human genesis toward naturalistic explanations that connected the races of the world to a shared organic origin with plants and animals. Scientific naturalists such as Thomas Huxley (1825–1895), Francis Galton (1822–1911), and John Lubbock (1834–1913) were central figures in championing this naturalistic model of the world.[14] The rise of secular modes of knowledge also had a tremendous impact on nineteenth-century theories of human history, sparking all sorts of new research programs. As an example, the positivism of the French philosopher Auguste Comte (1798–1857) transformed the ideas of many nineteenth-century researchers along secular lines.[15] In the case of human history theories, scholars such as Buckle, the philosopher of science Herbert Spencer (1820–1903), and the anthropologist, Edward Burnett Tylor (1832–1917), were indebted to Comte's ideas. All of them articulated developmental models that saw human civilization progressing toward a scientific worldview.[16]

Large-scale changes to print culture from the early nineteenth century onward connected new forms of knowledge to wider audiences and further transformed scientific understandings of humanity's past. As A. B. Van Riper, James Secord, Peter Kjaergaard, and numerous others have discussed elsewhere, books such as James Cowles Prichard's *Researches into the Physical History of Man* (1813), Robert Chambers's *Vestiges of the Natural History of Creation* (1844), Charles Lyell's *Antiquity of Man* (1863), Thomas Huxley's *Man's Place in Nature* (1863), John Lubbock's *Pre-Historic Times* (1865), Edward Burnett Tylor's *Primitive Culture* (1871), and Charles Darwin's *Descent of Man* (1871), all expanded human history beyond the traditional biblical time frame of six thousand years and opened up new questions about human origins.[17]

There were also other significant transformations occurring during the nineteenth century that affected the historicization of humans in the British sciences. George Stocking, Janet Browne, David Livingstone, Adrian Desmond and James Moore, and myself have all shown that voyages of exploration were equally important in changing scientific notions of human diversity and history.[18] As Europeans encountered different types of people living throughout the world, they struggled to explain how these different races fit into the story of Adam and Eve as the original progenitors of all humans. This opened up new questions about human origins. Within these "contact zones" (as Mary Louis Pratt has called them), Europeans created racial characteristics and human developmental histories by juxtaposing their own languages, customs, habits, traditions, and physical features against those of the indigenous populations.[19]

The emergence of alternative chronologies of human history in Africa and Asia were equally significant in producing further challenges to those scholars who continued to uphold the Judeo-Christian narrative of the Earth's past, with some of these diverging time lines pushing human existence beyond thirty thousand years.[20] Archaeological sites in English caves—such as those found near Torquay in Devon between the 1820s and 1860s—also generated important questions about human history, with skeletal and archaeological remains being discovered below the limestone strata. This discovery indicated that there was human life beyond the customary biblical time line.[21] All of these new types of evidence and theories brought the topic of human history to the fore of nineteenth-century British society. Within the context of British science and medicine, these discussions centered on the issue of human origins. What emerged from these heated dialogues were different types of explanatory models that aimed to describe human history and attempted to explicate the causes that created racial variation.

HISTORIOGRAPHICAL CONTEXT

Because the focus of this collection is on the historicization of humans in nineteenth-century British sciences there are complementary works in several historiographical areas. The following chapters build on major themes in the history of evolutionary studies by scholars such as Peter Bowler, James Elwick, Robert Kenny, and Gregory Radick, who have looked at various forms of developmentalism in the nineteenth century.[22] There are also strong thematic links to the works of Stephen Jay Gould, David Oldroyd, Ralph O'Connor, and Martin Rudwick, who have discussed in detail changing understandings of geological time in nineteenth-century scientific texts.[23] During the nineteenth century, topics such as race and empire were intricately tied to discussions of human history; illustrative examples can be drawn from the historiography on Victorian anthropology by George Stocking, Henrika Kuklick, Chris Manias, and Douglas Lorimer as well as the secondary literature on nineteenth-century British imperialism by Catherine Hall, Andrew Thompson, Sujit Sivasundaram, and Daniel Headrick.[24]

This volume moves beyond previous work on nineteenth-century human history, however, in three important ways. First, rather than looking at studies of human history through one discipline such as geology or paleontology, this collection will explore and cross-compare multiple disciplines, including geology, paleontology, natural history, archaeology, anthropology, and physiology. No single research field adequately represents nineteenth-century human history theories, because most researchers approached the topic through multiple disciplinary perspectives. Moreover, the boundaries between these disciplines were still being negotiated. There was tremendous overlap between the research programs of fields such as geology, geography, and paleontology as well as archaeology, anthropology, and history. A multidisciplinary approach, which considers the interconnections between various research fields, allows for a more nuanced examination of nineteenth-century human history theories.

Second, building on recent themes in imperial history, this book will take seriously the role of the colonial world in shaping nineteenth-century scientific understandings of human history. The aim is not to produce a complete global study. Instead, these chapters provide a selection of illustrative examples to show how imperialism shaped ideas on human history in various ways, depending on different colonial contexts. Correlations can be drawn between these different settings, while at the same time highlighting the distinct imperial conditions. Such an approach does not, therefore, necessitate a full coverage of every continent. Nevertheless, the volume does examine a broad range of geographical locations, including particular attention to Latin America, Canada, and South Asia.

A major reason for an emphasis on Latin America, featured in chapters 3 and 4 of the volume, is because its borders opened up to Britain after the collapse of the Spanish Empire at the beginning of the nineteenth century. This in turn brought Latin America to the forefront of British scientific and medical investigations. There were opportunities for British travelers of various shades to explore the area and collect new data, which transformed discussions on human origins and histories.[25] Canada, which is the focus of chapter 2, provides another interesting case study. Although Canada was firmly embedded within Britain's imperial network, the scientific research being carried out by some of its more high-profile figures was distinctly different from the approaches used by leading practitioners in Britain. This brings a key point to the fore: just because a particular method or theory was dominant in the metropole does not mean that it held the same importance in the peripheries. Science in the empire was diverse, and it was shaped by all sorts of local, national, and international influences. South Asia, which is examined in detail in chapter 5, was another significant location for Britain during the nineteenth century. Not only did it provide a wealth of materials and resources that fed the British economy at home and abroad, it also generated important scientific and medical information for researchers working in various disciplines (such as ethnology and anthropology) to utilize in their studies.[26]

As Catherine Hall and Sonya Rose have argued, you cannot fully investigate nineteenth-century Britain without considering its empire.[27] All aspects of British society during this period were affected by transformations occurring across the nation's vast imperial landscape. Every historical figure in this volume drew on Britain's imperial resources, whether they traveled abroad or stayed at home. Through the use of correspondence, travel reports, networks of colonial agents, or other means, the British Empire provided researchers with a wealth of material to incorporate into their investigations of humanity's past. There was also no monolithic conception of the British Empire during the nineteenth century. As Andrew Thompson has discussed in his book, *The Empire Strikes Back*, historians should start examining the pluralistic nature of the British Empire and consider how historical actors engaged with it differently depending on their personal circumstances.[28] Moreover, through a critique of the standard center-periphery model in British imperial studies, the essays in this collection will recognize the multi-directional nature of the traffic of ideas and influences between the metropole and colonial world.[29] In certain cases some of the essays will also cross-compare the British context with other European contexts. Some attention will be paid to nonspecialist understandings of human history through examinations of the general periodical press and other popular works.[30]

Third, the essays in this collection will look at the various forms of human developmental theories that were competing for scientific dominance throughout the British Empire in the nineteenth century. Darwinism will feature in some of them, but there will also be detailed analyses of other developmental theories from the nineteenth century—such as those derived from disciplines as diverse as philology and embryology. Building on Bowler's concept of the "non-Darwinian revolution," these chapters will push beyond the standard historiographical narrative that has prioritized Darwinian evolution, to show that there were other significant developmental models transforming human history theories in the nineteenth century.[31] All three of these historiographical points are not in themselves novel. However, by bringing these three disparate bodies of secondary literature into conversation and emphasizing important intersections, it is possible to construct a new historiographical narrative that deepens our understanding of human history theories in the nineteenth century, showing the subtleties and nuances that existed.

HISTORICIZING HUMANS THROUGH CASE STUDIES

In his 1992 essay "Retrospective Prescriptive Reflections," the historian George Stocking discussed how he often favored an approach that examined different "vignettes" that occurred during the disciplinary history of anthropology, because it allowed him to examine the "multiple contextualizations" of anthropology's past. This historiographical method is particularly useful for a collection of case studies such as this one, as it affords an opportunity to cross-compare how a diverse group of historical actors from different geographical, disciplinary, and social contexts historicized humans. Despite these distinctions there were some underlying questions that pervaded all of these nineteenth-century practitioners' research programs: What was the age of the Earth? Where and when did humans first appear? Who had the authority to speak about human history? Should priority be given to science or religion when discussing humanity's past? What types of data should be used by researchers when studying human history?

Our study begins in the deep past, when humans lived among extraordinary animals such as woolly rhinoceroses. As Chris Manias discusses in chapter 1, the middle of the nineteenth century witnessed a time revolution, where naturalists were bursting the boundaries of human history. The races of the world were thrust back into prehistoric environments with strange and long-extinct beasts. These shifting conceptions of a deeper human history conflicted with traditional, religiously influenced understandings of the natural world and human-animal relations. As we will see in due course, scientific and medical figures responding to these temporal changes in human history began reevaluating humanity's power over nature, and specifically, its interaction with prehistoric fauna. Manias argues that this opened up a series of new questions: What was the relationship between prehistoric humans and animals? How could primitive people survive among such fierce creatures? Were humans ordained with a special mastery over the organic world, or was there some evolutionary mechanism at work? What role did prehistoric humans play in the disappearance of these ancient animals? Whatever the results of these reconsiderations of human time lines were, one thing was clear: that human-animal relations were historicized in hitherto unknown ways.

The connection between science, religion, and human history is further brought to the fore in the next two chapters. In chapter 2, Nanna Kaalund looks at *Archaia* (1860), the first popular work by the Canadian geologist, paleontologist, and university administrator John William Dawson (1820–1899). *Archaia* was a monogenetic study that attempted to harmonize biblical and scientific narratives of human origins. As we will see in due course, Dawson proposed a day-age theory, where each of the seven days of creation marked a geological period of time. Kaalund explores Dawson's double commitment to science and religion, examining the types of strategies that he used to reconcile these two spheres in his writings from *Archaia*. Dawson was also a chief popularizer of science in Canada during the nineteenth century, and as Kaalund shows, he used his position as principal of McGill University to forward his particular vision of science—especially when it came to historicizing humans. Moreover, the Canadian context provides an interesting case study for showing differences between various scientific locations throughout the empire. Dawson's particular form of natural theology, for example, received far less criticism in Canada than in Britain, because scientific naturalism had less of a hold there.

In chapter 3 I look at the developmental writings of the ethnologist-turned-anthropologist Edward Burnett Tylor. In particular, I explore how Tylor historicized religious beliefs in his magnum opus, *Primitive Culture*. Through a detailed examination of this work, I discuss how Tylor constructed his theory of animism by exploring the various influences that shaped his writings both within Britain and throughout the empire. In contrast to Dawson, who tried to harmonize science and religion, Tylor wanted to naturalize all religions and explain their ontologies using scientific theories. He was not trying to reconcile science and religion but, rather, to bring religion under the domain of scientific understanding. However, despite Tylor's clear aim to replace religious explanations of the world with scientific ones, a close examination of Tylor's book exposes a complex and strenuous relationship between science and religion. Many of the sources that he used to exemplify extra-European religious practices came from the travel reports of missionaries and other types of travelers with strong religious convictions. Tylor was reliant on these firsthand accounts to substantiate the credibility of his writings. Even though he wanted to push religion to the margins of ethnological and anthropological research, he was unable to avoid religiously influenced sources for his data completely. This raises some interesting questions about the alleged boundaries between the so-called two spheres of science and religion.

The influence of the British Empire in transforming understandings of human history is a major theme in the next two chapters of the volume. As Maurizio Esposito and Abigail Nieves Delgado show in chapter 4, among the many nineteenth-century scientific figures to find employment through travel was the British natural history collector and anthropologist William Bollaert (1807–1876). Though largely forgotten among English-speaking scholars today, Bollaert was one of the leading mid-nineteenth-century experts on South American indigenous peoples. His work, which historicized the races of the world through a polygenetic framework, was respected among members of the British anthropological community, and he presented much of his research at the Anthropological Society of London (ASL) in the 1860s, with several of his papers being published through the ASL's periodicals. Nevertheless, because Bollaert's primary income derived from natural history collecting, his investigations of South American peoples were largely motivated by monetary concerns. Esposito and Nieves Delgado, therefore, examine the significance of imperial networks of exchange in shaping Bollaert's anthropological writings.

In chapter 5, Thomas Simpson moves the focus further abroad to India, and he underscores the significance of the imperial periphery in shaping knowledge about human history. Two key themes to emerge in this chapter are those of authority and power. Who had the authority and power to construct conceptions of human history—especially in the colonial world? Simpson explores these issues by looking at geographies of knowledge. Simpson discusses how in colonial India, arguments over human origins and its diversity were distinctly different from similar debates occurring in metropolitan Europe. That is not to say efforts to historicize humans in India were isolated from those happening in Britain and elsewhere but, rather, that the specific colonial context had a direct bearing on the nature of the debates. For instance, as Simpson shows, internal squabbles between the main Indian centers of Calcutta, Madras, and Bombay played an important factor in local discussions of humanity's past, with each locale fighting for cultural hegemony in the region.

Efforts at tracing deeper human histories were a mainstay in nineteenth-century scientific studies of human origins, and these newer conceptions of time included examinations of not only the physical attributes of races but nonphysical ones as well. This aspect of human history theories is central to Gregory Radick's analysis in chapter 6, where he looks at Charles Darwin's attempt at historicizing human feelings in *The Expression of Emotions in Man and Animals* (1872). As Radick shows, Darwin's preoccupation with trying to establish a common origin for all humans was intricately linked to his views on slavery, and much of his writing on human evolution occurred in the wake of the American Civil War, when debates about monogenesis and polygenesis were coming to a head. This argument is well known among historians of Darwin, but Radick argues that the strongest evidence for it does not lie in the *Origin of Species* or the *Descent of Man*, as previously articulated in the historiography, but in *Expression of Emotions*. Furthermore, Radick's discussion reminds us that sociopolitical factors had a strong impact on how human histories were constructed in the nineteenth century, with claims of ideological neutrality being part of the vocational strategies of Victorian scientists.

Not all conceptions of human history in the nineteenth century looked at deep time, and in chapter 7, Helen Kingstone explores the theme of shallow time in the works of the historical novelist and playwright Walter Scott (1771–1832); the philosopher, essayist, and historian Thomas Carlyle (1795–1881); and the scientific polymath Francis Galton. Even when tracing shorter time periods, historical records could be sparse, and in dealing with this issue researchers relied on conjectural methods to fill in gaps. These methodological techniques were highly flexible, could be applied across short or long time frames, and were easily appropriated into either monogenetic or polygenetic frameworks. The comparative method was particularly advantageous for researchers attempting to connect disparate human groups across time and space, according to their levels of civilization. Studies of shallow time (with its commitments to conjectural and comparative methods) also could be useful in elucidating deeper understandings of the past; not merely in accounting for the transition from rude to civilized societies but also for understanding evolutionary and geological shifts in the natural world. Nevertheless, as Kingstone emphasizes in her study, there was another element in these historical frameworks, one that attempted to predict where more primitive civilizations were heading, based on the histories of European societies. This sort of futurism was present in the works of many Victorian researchers interested in the historicization of humans.

In Chapter 8, Ian Hesketh further examines in overview the role of futurism in the works of a diverse cast of evolutionary writers from the 1860s to the early 1900s. His study includes the writings of Alfred Russel Wallace (1823–1913); Galton; the British explorer and anthropologist W. Winwood Reade (1838–1875); the classicist, poet, and psychical investigator Frederic Myers (1843–1901); the zoologist E. Ray Lankester (1847–1929); and the evolutionist and spiritualist John Page Hopps (1834–1911). Hesketh argues that, even though the religious views of these evolutionists ranged widely from agnostic to spiritual in nature, they all depended on notions of Christian eschatology in order to instill their evolutionary narratives with grand cosmic meaning. Thus, when it came to historicizing humans in the works of these evolutionary writers, the intersection of science and religion comes back into the frame, full circle.

In the afterword, Theodore Koditscheck provides some historiographical reflections on the significance of this collection for the larger secondary literature. Most notably, by drawing together the major themes and arguments from the various chapters, he shows how each author in this volume expands the standard historiographical narrative on nineteenth-century human history theories in hitherto unknown or overlooked ways. Koditschek divides the secondary literature into two camps: "Historiography of Historicizing Human Origins 1.0," which is represented by the older guard of scholars who pioneered studies on nineteenth-century human history theories, and "Historiography of Historicizing Human Origins 2.0," which is represented by the newer generation of scholars who are broadening the analytical landscape. Taking them together, he envisions a bright future for the research field. With all of these historiographical themes in mind, let us now direct our attention toward the eight case studies, and consider how each chapter examines differently the historicization of humans in nineteenth-century British sciences.

1

CONTEMPORARIES OF THE CAVE BEAR AND THE WOOLLY RHINOCEROS

Historicizing Prehistoric Humans and Extinct Beasts, 1859–1914

CHRIS MANIAS

In his popular lectures and articles on "the British Lion" delivered at various points over the 1880s, the Manchester geologist William Boyd Dawkins (1837–1929) presented primeval Britain as a dark and terrifying place. Synthesizing data from the new fields of prehistoric archaeology, which had unveiled human existence deep in geological time, and paleontological discoveries of the bones of large herbivores and predators, he created a vignette of life in the ancient Thames Valley, when it was "haunted by many extinct wild animals, and by living species no longer found together in any part of the world." Familiar creatures such as "stags and roe-deer lived in the forest side by side with the gigantic and extinct Irish elk, the woolly rhinoceros, and the straight-tusked elephant," not to mention "innumerable horses, bisons, and large horned uri." There was a more complex and potentially more dangerous element: "these animals were kept in check by numerous beasts of prey; the smaller of them by stealthy foxes and wild-cats, and the larger by grizzly and brown bears and packs of wolves. The stillness of night was from time to time broken by the weird laughter of the spotted hyena and by the roar that proclaimed the presence of the king of beasts'—the titular 'British Lion.'" But within this, humans were a key presence:

> The central figure, however, in the picture is proved by recent discoveries to have been man. Not only have flint implements of the ordinary river-drift type been obtained from the brick-earths of Crayford along with remains of the animals above mentioned, but Mr. Flaxman Spurrell has been able to fix the place where the hunter sat on the ancient bank of the Thames and fashioned the blocks of flint to his various needs. The river-drift hunter, armed with his roughly chipped stone implements, doubtless had great difficulty in making good his place in the struggle for existence among the beasts of prey then in the valley of the Thames, and sometimes, when he had the chance, he would be likely to eat the lion, and at other times the lion would certainly eat him. They must often have come into contact when engaged in the pursuit of the same animals.[1]

The strangeness and fierceness of the fauna living alongside prehistoric humans raised a number of issues for Victorian thinkers. How had the most primitive humans survived alongside some of nature's most fearsome creations? While few doubted that humans were ordained to mastery over the natural world, the findings of prehistoric archaeology and paleontology made these relationships difficult to understand.

The "establishment of human antiquity" and the entrenchment of a concept of human prehistory from the late 1850s was one of the most dramatic developments in the historicization of humans in the nineteenth-century sciences. A growing historical literature is now asserting that this mid-Victorian "time revolution" was as significant for understandings of "man's place in nature" as the much better studied "Darwinian revolution."[2] This literature has traced how new approaches to human prehistory synthesized disciplinary perspectives from geology, archaeology, paleontology, natural history, and anthropology and worked across a range of geographic boundaries, requiring collaboration from scholars across Europe and beyond.[3] A new series of geological, paleontological, and archaeological eras were slowly elaborated and interrelated to one another.[4] As absolute dates were impossible to deduce, these relative eras were consistently understood as showing increasing complexity—and often "progress"—in human development. Reflections on human prehistory were therefore highly significant for key themes in nineteenth-century intellectual and cultural life, including concepts of race, social development, "savagery," gender, and religion, which were all interrelated and altered when understood within the deep developmental chronologies of human prehistory.[5]

One of the key disciplinary links in these processes was between paleontology and the archaeology of the European Stone Age, with the former being the more established science. That Britain and other parts of Europe in the Pleistocene period and Ice Age had been inhabited by such creatures as hyenas, mammoths, woolly rhinoceroses, hippopotamuses, cave bears, and lions were some of the first discoveries of the deep time sciences in the eighteenth and nineteenth centuries, predating the establishment of human antiquity by many decades.[6] The key initial proof in the "establishment of human antiquity" was the discovery of human stone tools in the same geological strata as the remains of these extinct animals—meaning that their contemporaneousness was enshrined in the methods of prehistoric research.[7]

The secondary literature has tended to emphasize these more practical dimensions, with paleontology supplying the raw evidence to establish human antiquity. However, awareness of the coexistence of humans with these extinct beasts was also significant on more conceptual levels, as it raised a range of questions on early humanity's relationship with nature. What had been the relationship of early humans with these large and strange animals? How had Stone Age peoples not only survived but apparently triumphed over such a dangerous fauna? And what role (if any) had humans played in the disappearance of creatures such as the mammoth, the woolly rhinoceros, and the great cave bear? In discussing these questions, Victorian commentators were forced not only to seek analogies in the modern world but also to reflect on humanity's place in nature or (for more religious writers) the divine order of things, reflecting on how far the prehistoric development of humans had been a process of triumph over, conflict with, or escape from a primordial nature.

This chapter will follow how nineteenth-century thinkers conceptualized the connections between humans and prehistoric beasts. The first section will examine the early development of the field, how human prehistory and Pleistocene paleontology were linked through the coexistence of humans and large Ice Age mammals. The second section will consider how this material evidence was used to reconstruct the relationships between humans and the herbivores, and predators of Ice Age Europe, particularly discussing how this fed into ideas of the capacities and early development of "primitive man." The final section will consider how the ultimate fate of the animals of prehistory, which had either become extinct or migrated away from Europe at the dawn of the historic period, was conceptualized. Across these areas, human interaction with the beasts of prehistory was generally understood in terms of transition, providence, and human domination over the natural world—a transition from "The Age of Mammals" to "The Age of Man." However, the lessons that were drawn from this were highly variable, and the tone adopted could veer from confident triumphalism to melancholic reflection.

AROUND THE ESTABLISHMENT

The animals of the Pleistocene had generated sensational interest in late eighteenth and early nineteenth-century Europe. This was partly due to scholars such as the French comparative anatomist Georges Cuvier (1769–183), making dramatic public lectures to large audiences, or William Buckland (1784–1856) in Britain, venturing into caves to bring back the remains of extinct animals to present to interested scholars. Many other remains were unearthed in the course of mining, road building, and urban construction projects. As these finds were assembled and correlated, they showed a relatively recent prehistoric environment inhabited by elephants, rhinoceroses, hippos, hyenas, lions, giant deer, and bison. Paleontological discoveries from outside of Europe accentuated a sense of strangeness in the deep animal past: the giant ground sloths and glyptodonts of South America were extensively discussed as bizarre aberrations;[8] the North American interior delivered up a further suite of mammoths, mastodons, and saber-toothed cats from the 1840s onwards; and the (often highly fragmentary) remains of giant fossil marsupials from Australia were received in Britain and widely discussed.[9] Prehistoric faunas were understood as varied, diverse, and strange, and the finds were interpreted within a wide global context.

Ralph O'Connor has argued that the more recent remains of the European

Pleistocene were overshadowed and regarded as "old news" as the nineteenth century moved on and even more spectacular discoveries of fossil marine reptiles and dinosaurs began to attract attention, but there was still a strong and continual interest throughout the period in Pleistocene Europe.[10] The fauna of the period was understood as highly diverse, apparently mixing animals from different modern biogeographic zones, including Europe, Africa, and Asia. This illustrated an impressive and varied fauna and also implied a shifting climate. Some of the animals, such as the reindeer and mammoth, were clearly adapted for cold—even Arctic—conditions, while others, such as the hippo, were more suited for warmer environments. Understandings of the epochs prior to the Pleistocene—the Pliocene and the Miocene—accentuated these ideas of climatic shifts even further: these periods were presented as lush and warm, with Europe potentially covered in tropical jungles and hot savannahs, filled with apes, gazelles, saber-tooth cats like the *Machairodus*, and a variety of forms of elephant and mastodon. This raised the implication of dramatic environmental change, primarily manifesting as gradually cooling down over the ages.

In the period before the establishment of human antiquity, these ideas—of climatic difference and the strangeness of the fauna—were often used as evidence against human existence in deep geological eras. This could be achieved in a variety of ways, often by the same author. Some possibilities can be seen in how the doyen of mid-Victorian British paleontology, Richard Owen (1804–1892), interpreted two types of creature. In his survey of 1846 on *British Fossil Mammalia*, he expressly made the comparative analogy between modern wild carnivores outside of Europe and their prehistoric counterparts:

> When we are informed that, in some districts of India, entire villages have been depopulated by the destructive incursions of a single species of large Feline animal, the Tiger, it is hardly conceivable that Man, in an early and rude condition of society, could have resisted the attacks of the more formidable Tiger, Bear, and *Machairodus* of the cave epoch. And this consideration may lead us the more readily to receive the negative evidence of the absence of well-authenticated human fossil remains, and to conclude that Man did not exist in the land which was ravaged simultaneously by three such formidable Carnivora, aided in their work of destruction by troops of savage Hyaenas.[11]

Early humans could not have successfully engaged in a struggle for life with such dangerous predators, and human survival required the disappearance of the large carnivores. The assumed ferocity of the prehistoric fauna could be used as proof against human existence.

However, it was not just that prehistoric carnivores were conceptualized as being too dangerous for prehistoric humans. Some specimens were interpreted in the opposite manner, with humans thought to have been too powerful a predator for the animals to survive alongside. Owen's 1842 analysis of the *Mylodon robustus*—a South American ground sloth—raised this issue directly. Much of the monograph reconstructed the animal's habits and lifestyle and devoted a great deal of attention to interpreting a partly healed wound on the animal's head. Owen dismissed various possibilities for what could have caused this, including potential human predation, noting:

> There is no certain or conclusive evidence that Human Beings co-existed with the Megatherian animals; but assuming a primeval race of Indians to have disputed the lordship of the American forests with the Edentate giants, and to have waged against them, as against all other inferior animals, a war of extermination; the same difficulty presents itself to the supposition of the recovery and escape of a stunned *Mylodon* from their deadly assaults with clubs and other weapons, as from the claws and teeth of the beast of prey: for the flesh of the leaf-eating Megatherian would doubtless be as much prized for food by a Human destroyer as that of the Sloth is by the Indians of the present day.[12]

This passage indicates important aspects of conceptualized human interaction with large herbivores. Humans and large beasts were anathema to one another, and large herbivores would not have been able to survive conflict with human hunters—even when the latter lived in a "savage" state. In this way, human interaction with animals could be understood in quite different terms, depending on the animal's place within the "economy of nature."

These modes of reasoning and related doubts on applying deep time chronologies to human existence ensured that in the few early nineteenth-century cases where human remains or tools were discovered alongside prehistoric animals, judgments were strongly against them being contemporary. Famously, William Buckland denied that the human skeleton named the

"Red Lady of Paviland" found in Wales in the 1820s was the same age as the mammoth and other prehistoric animal remains found alongside it; instead he argued it was a later interment.[13] While Buckland argued this partly from scriptural grounds, other writers interpreted finds in other ways. In 1832 the Belgian priest Philippe Schmerling (1791–1836) discovered a human skull alongside the remains of hyenas, lions, and other animals and judged them as coeval, but these conclusions were largely ignored or rejected by contemporaries—although Schmerling's finds did later become iconic as prehistoric human remains.[14] Partly this was because of conceptual reluctance to accept human antiquity. However, it was also because of the nature of the cave sites being investigated, which contained skeletons strewn around cave floors without any clear indication of the time when the individual remains had been deposited.

The manner in which these evidences were drawn together has been discussed in a number of works on the "establishment of human antiquity."[15] In the late 1850s a number of human tools discovered in ancient geological strata were authenticated, particularly those found by the French antiquary Jacques Boucher de Crèvecœur de Perthes (1788–1868) in the Somme valley, and then scrupulous excavations of a series of prehistoric sites, particularly in southwest England. Localities such as Brixham Cave near Torquay contained human tools buried in the same ancient geological strata alongside the remains of prehistoric animals. These discoveries were promoted in a systematic manner to the leading metropolitan learned societies in London and Paris in the years 1859–1860 and were followed by further studies and popularizing works. In all of these presentations, evidence that humans had lived alongside the Pleistocene animals was crucial. The existence of the prehistoric mammals was well established in scholarly and educated circles by this period, and so this was very much a slotting of humans into already set categories.

Yet the connection also made the newly elaborated field of human prehistory more dramatic—Charles Lyell was to open his work *Geological Evidences on the Antiquity of Man* with the lines: "No subject has lately excited more curiosity and general interest among geologists and the public than the question of the Antiquity of the Human Race—whether or not we have sufficient evidence in caves, or in the superficial deposits commonly called drift or 'diluvium,' to prove the former co-existence of man with certain extinct mammalia."[16] Depicting the finds in terms of "that portion of the post-Pliocene period in which Man co-existed in Europe with the mammoth" meant that the earliest humans were immediately pictured in terms of their interaction with these mighty animals.[17] It also led to reflections (presaged by Richard Owen's above doubts) on how humans could have lived alongside—and competed with—such creatures. Human prehistory not only used the animals as scientific evidence but posed human-animal interaction as a talking point and a conceptual problem.

Further sites were interpreted even more clearly through the framework of human-animal interactions. Some of the most striking were the remains discovered at Cro-Magnon in the Dordogne by Édouard Lartet (1801–1871) and Henry Christy (1810–1865) in the 1860s, eventually published in a sumptuous volume titled *Reliquiae Aquitanicae* (1875).[18] The caves were interpreted as habitation sites, which included human remains and "multitudinous examples of bones, broken up by man, of animals extinct in that part of Europe, out of all record of history or tradition."[19] This assemblage allowed a reconstruction of prehistoric diet and hunting practices, with most of the bones being of medium-sized mammals such as reindeer, aurochs, and horses. Moreover, the caves also contained artistically carved bone, antler, and ivory, which depicted the whole range of Pleistocene fauna, including mammoth, wolverine, reindeer, horse, and various bovids. These provided direct evidence of human observation of these creatures and showed the people of Cro-Magnon as intensely interested in the fauna around them.

Nineteenth-century French accounts of human prehistory were in some respects even more strongly predicated upon linking humans with animals than British accounts. Lartet, for example, divided the earliest phases of human prehistory into successive Cave Bear, Reindeer, and Aurochs ages, each defined by the predominance of a particular creature.[20] These terms were not limited to scientific texts and formed the organizing principles for popularizing works, most notably Louis Figuier's *L'Homme Primitif*, published in 1870 and translated almost immediately into English.[21] However, the animal ages were not widely accepted in Britain (and were opposed later in France), where Christian Thomsen's material ages of Stone, Bronze, and Iron became preferred over the 1860s, at least in prehistoric contexts.[22] This was partly because of a more materialist and civilizational bent that prevailed in British prehistory, but it was also because of arguments on the diversity of the prehistoric fauna. Boyd Dawkins was to write that "it is easy to refer a given cave to the age of the reindeer or of the mammoth because it contains the remains of those animals, but the division has been rendered worthless for chronological purposes, by the fact that both these animals inhabited the region north of the Alps and Pyrenees at the same time." Selection bias was a more likely cause, as "the abundance of the reindeer, which is supposed to characterize the reindeer period, may reasonably be accounted for by the fact, that it would be more easily captured by a savage hunter, than the mammoth, woolly rhinoceros, cave-bear, lion or hyaena."[23] Early human interaction with prehistoric animals was complex and depended on the character of the fauna, available technologies, and the capabilities of early humans to hunt particular animals. Interpreting the contents of the caves therefore depended on an understanding of early human interaction with the animals and built on assumptions of human capacities and levels of development.

INTERACTIONS

The very "establishment of human antiquity" was based upon evidence that humans and prehistoric animals had coexisted and interacted. However, once this had been ascertained, different questions were posed: What had been the nature of this interaction? Were prehistoric humans the masters of the prehistoric environment, or had they eked out an existence on the margins? While Victorian thinkers had little doubt that modern industrial civilization was at the pinnacle of creation, and that humans were on a different plane from other animals, exactly when this superiority had become manifest was not agreed on at all and formed the root of many of the discussions of human interaction with prehistoric beasts. Whether justified through notions of progressive development or through Christian ideas of humans being given mastery over the Earth and all the creatures in it, the actual point at which human dominance had occurred was difficult to pinpoint. Whether the Stone Age represented the beginning of human mastery or a threatened preexistence before it was a central issue to be worked out in discussions and reconstructions.

These reflections were also affected by understandings of variation in the natural world. The fauna of prehistoric Europe was judged to have been extremely diverse, both in terms of the range of animals fulfilling specific roles within the ancient "economy of nature" and because the differing eras of prehistory had different distributions of particular animals (something already presaged in Lartet's paleontological ages).[24] Humans were also felt to have had different capabilities in relation to different types of animals, with interaction between large and dangerous carnivores (such as the cave lion, cave bear, and hyena), giant herbivores (such as rhino, mammoth, and hippo), medium-sized herbivores (including reindeer, aurochs, and horses), and smaller animals (such as beavers, voles, and foxes), all being seen as taking different forms. Some animals could be regarded as prey, some as rivals, and some as potentially dangerous but not necessarily in conflict with prehistoric humans.

It was not only the case that the animals of prehistoric Europe were understood as having been highly diverse but the prehistoric environment was also understood as having seen a great deal of climatic and environmental change. The general models within paleontology tended toward a steadily cooling climate, with the lush jungles and hot savannahs of the Miocene giving way to the warm plains of the Pliocene and then the harsh cold of the Pleistocene—which was in turn followed by a slight warming of the climate in more recent times. A further complicating factor arose toward the end of the nineteenth century, when the concept of a single glacial period in the Pleistocene Ice Age began to fade and was replaced by a model of a succession of freezings and warmings, and glacial and interglacial periods. All this ensured that humans and animals were placed within a dynamic climatic system that changed over these eras of Earth history. Pinpointing exactly when humans entered this changing environment was therefore an important topic of debate.

Attempts to understand the interaction of humans with prehistoric animals therefore took a wide range of factors into account. Nevertheless, some writers and commentators depicted these developments as a fairly steady and linear story of the growth of human supremacy. Notably, these presentations were often derived from (or were at least highly influenced by) the French context, where ideas of linear development were strongly presented.[25] Louis Figuier's *L'homme primitif* followed this model quite specifically: his depictions showed early humans fighting the cave bear and hyena over habitation sites, then moving to hunt reindeer in the forests on foot, and finally (in the Bronze Age) domesticating horses to aid in the hunting of elk and developing agriculture supported by domesticated goats and cattle. The early stages were specifically cited as difficult and dangerous. Figuier opened his work by highlighting the threatening nature of the animal life, which early humans had faced:

> The mammoth, elephant, rhinoceros, stag, and hippopotamus were then in the habit of roaming over Europe in immense herds, just as some of these animals still do in the interior of Africa. These animals must have had their favourite haunts—spots where they assembled together in thousands; or else it would be difficult to account for the countless numbers of bones which are found accumulated at the same spot.
>
> Before these formidable bands, man could dream of nothing but flight. It was only with some isolated animal that he could dare to engage in a more or less unequal conflict. Farther on in our work, we shall see how he began to fabricate some rough weapons, with a view of attacking his mighty enemies.[26]

The rest of Figuier's book showed steady human invention of technologies, which gave power over the natural world. The first of these was fire, which aided the exploitation of habitation sites and allowed the consolidation of familial bands and social groups. This collective life meant that "man felt the need of strengthening his natural powers against the attacks of wild beasts. At the same time he desired to be able to make his prey some of the more peaceable animals, such as the stag, the smaller kinds of ruminants, and the horse. Then it was that he began to manufacture weapons," and so, "at the epoch of the great bear and mammoth, was able to repulse the attacks of the ferocious animals which prowled round his retreat and often assailed him." Animal body parts were also used in human technology, and "reindeer's horn was the earliest raw material in the manufactures of these remote ages, and to the man of this epoch was all that iron is to us."[27] This showed steady human control over the natural world and—after fearful beginnings—turned into a story of progress and mastery.

This narrative of progressive human supremacy was also presented in one of the most popular accounts of human prehistory, John Lubbock's *Pre-Historic*

Fig. 16. L'homme à l'époque du grand ours et du mammouth.

FIGURE 1.1. "Man during the Age of the Great Bear and Mammoth." From Louis Figuier, *L'homme primitif* (Paris: Librairie de L. Hachette, 1870), 53. Author's collection.

Fig. 54. La chasse à l'epoque du renne.

FIGURE 1.2. "The Hunt During the Reindeer Age." From Louis Figuier, *L'homme primitif* (Paris: Librairie de L. Hachette, 1870), 113. Author's collection.

Times (1865), a work that simultaneously gave an account of the recent discoveries in human antiquity and reconstructed the lives and habits of humans in different eras through a combination of archaeological summaries and ethnological analogies, with a particularly important role being played by the "comparative method"—the idea that modern "savages" and prehistoric Europeans were analogous. A great deal of attention was paid to human interactions with the animals of the prehistoric epochs, with Lubbock giving an account of the "reindeer hunters" described by Lartet and Christy, the shellfish-eating inhabitants of the Danish coast, and the lake-dwellers of Switzerland. Lubbock's comparative analogies, however, made him much more confident of the power of prehistoric hunters over their animal rivals. He noted how travelers described the "Hottentots" of South Africa as able to hunt large African animals with spears, bows, and javelins: "With these weapons they were very skilled, and feared not to attack the elephant, the rhinoceros, or even the lion. Large animals were also sometimes killed in pitfalls, from six to eight feet deep, and about four feet in diameter. They fixed a strong pointed stake in the middle. 'Into this hole an elephant falling with his fore-feet (it is not of dimensions to receive his whole body) he is pierced in the neck and breast with the stake and there held securely,' for the more he struggled the farther it penetrated."[28] This use of ethnographic parallels and assumption that the "Hottentots" (often judged to be at the same level of civilization as the Old Stone Age inhabitants of Europe) could hunt elephants not only illustrated that prehistoric Europeans could potentially have hunted large creatures like the mammoth and woolly rhinoceros but also gave an indication of the kind of hunting strategy they could have used. Lubbock saw human "savages" as immoral, irrational, and childlike in relation to modern "civilized" humans, but in terms of their relations with the animal world they were far superior. If "the North American Indian will send an arrow right through a horse, or even a buffalo . . . the African savage will kill the elephant, and the Chinook fears not to attack even the whale,"[29] Ice Age Europeans were likely similarly fearless and effective.

Lubbock's ideas on ancient human hunting were dramatized in a series of privately commissioned paintings by the artist Ernest Griset (1844–1907).[30] These nineteen paintings were hung in Lubbock's estate and showed prehistoric Europeans in a variety of settings and poses. Many of these depicted

FIGURE 1.3. "Mammoth Hunt." By Ernest Griset (1860s/1870s). Courtesy of Bromley Historic Collections and Professor Tim Murray.

FIGURE 1.4. "Bison Hunt." By Ernest Griset (1860s/1870s). Courtesy of Bromley Historic Collections and Professor Tim Murray.

FIGURE 1.5. "Cavern Scene." By Ernest Griset (1860s/1870s). Courtesy of Bromley Historic Collections and Professor Tim Murray.

interaction—always violent—with prehistoric animals. Ancient humans were shown hunting stag and bison in two separate paintings, and a further image dramatized a prehistoric hearth in a cave, with a dead reindeer lying on the floor. A further image, echoing that of Louis Figuier, depicted humans fighting a cave bear over a habitation site, with one human about to be mauled, while the other is ready to strike a strong blow. The most dramatic image—"the mammoth hunt"—showed a large band of humans throwing spears at an angry mammoth, with numerous dead hunters scattered about. Clearly it was a highly dangerous activity. These images showed human dominance over all parts of the natural world, with cooperative hunting allowing triumph over medium-sized herbivores, great carnivores, and even giant herbivores like the mammoth, although with varying degrees of risk and success. Social organization in the hunting and savage state was therefore key for human superiority.

These notions of growing triumph in the Old Stone Age were not the only possible interpretations, however, and many writers took almost the opposite view, emphasizing humanity's weakness and apparent defenselessness against the fearsome beasts of prehistory. Worthington George Smith's *Man, the Pri-*

meval Savage (1894) summarized a variety of finds in southeastern England and presented something of a counternarrative to the model of human dominance. He argued that the "primeval savages" were marginal inhabitants of prehistoric Europe and could not compete with the larger animals. He noted specifically: "Primeval man is commonly described as a hunter of the great hairy mammoth, of the bear and the lion, but it is in the highest degree improbable that the human savage ever hunted animals much larger than the hare, the rabbit, and the rat. Man was probably the hunted rather than the hunter."[31] This idea was further elaborated with a long exposition on how early humans would have interacted with the whole range of Pleistocene fauna. The ancient hippo of England's rivers, "not being a flesh-eating animal . . . would not be much dreaded by its human companions; the old bulls would, however, sometimes scatter human companies. Neither would the hairy mammoth and the straight-tusked elephant molest the men further than by an occasional charge from a furious old bull." Predators were more dangerous, as "the stealthy and terrible lion, silent and swift of foot, together with the spiteful and ferocious wild cat, would always strike terror into the heart of the primeval savage," and "the cowardly and terrible hyaena would frequently chase or pounce on men, women, and children . . . it would at times stealthily discover, bite, tear, and kill members of the human family at night."[32] These primeval Europeans were at the mercy of their environment and lived a life under constant threat. While Smith was clear that early humans still "represented the highest stage of development of the animal kingdom of [their] time," this was not because of their power and strength.[33] It was because of their possession of the rudiments of language, spiritual belief, society, and technology. However, these cultural attributes required a very long process of development to reach fruition, and this potential was by no means obvious in the depths of the Pleistocene.

While Smith held to a broadly evolutionary model, for other writers ideas of original human weakness were actually used to discount concepts of human evolution and to present matters in a more theologically inspired manner. In *The Story of Earth and Man* (1873), the Canadian geologist John William Dawson (1820–1899) drew attention to human physical weakness, particularly in comparison with apes, to argue that early humans could not have developed alongside the more dangerous animals of European prehistory.[34] Had they originated in such a dangerous environment and survived Darwinian competition with the natural world, humans would have been distorted by evolution to "become in their structures more like carnivorous beasts than men."[35] Instead, Dawson felt that early humans must have been sheltered away from ferocious animals through most of their origins, concluding that

> the only relational hypothesis of human origin in the present stage of our knowledge of this subject is, that man must have been produced under some circumstances in which animal food was not necessary to him, in which he was exempt from the attacks of the more formidable animals, and in less need of protection from the inclemency of the weather than is the case with any modern apes; and that his life as a hunter and warrior began after he had by his knowledge and skill secured to himself the means of subduing nature by force and cunning. This implies that man was from the first a rational being, capable of understanding nature, and it accords much more nearly with the old story of Eden in the book of Genesis, than with any modern theories of evolution.[36]

The development of humanity required peace, shelter, and isolation from dangerous nature, and it was only after humans were fully formed, and equipped with reason and technology, that they could dominate the natural world. This was a religiously derived view, attempting to reconcile paleontological findings with scriptural accounts. However, it is interesting to note that it was expressed within the same idiom as the other discussions of early human interaction with the beasts of prehistory and engaged with the same central problem: how early humans could survive alongside such dangerous animals. Yet here the question was interpreted through a scriptural framework (and answered in the negative) rather than an evolutionary one.

A number of highly synthetic works, which aimed to cover the whole of evolutionary history, also focused attention on humanity's original place within the natural world and interaction with extinct animals. A particularly notable example was the narratives produced by Henry Knipe (1854–1918), which included *From Nebula to Man*, an evolutionary epic in verse of 1905, and *Evolution in the Past*, a more standard text of the history of life appearing in 1912. These works traced the whole history of creation but placed a special emphasis on the rise of humans, mixing in new finds in paleoanthropology, particularly recently discovered earlier human types such as *Pithecanthropus* in Java and Neanderthals in Europe. In Knipe's works, human capacities to dominate the environment were presented as having been quite different at varying stages of human evolution. The earliest known human form, *Pithecanthropus*, was presented as being similar to Smith's primeval savages, primarily eating "small mammals, eggs, roots, and berries. Big game certainly haunted his neighbourhood—elephants, rhinoceroses, and hippopotamuses. . . . But it is doubtful if he had the wit or the power to overcome any of these animals."[37] It was only later in evolutionary history that humans could manage such a task. Later humans emerged into "a happy hunting-ground" in Glacial Europe, although it was still the case that "many of the animals were probably too formidable for him or too fleet of foot. . . . Man [was] destined to kingship, but as yet far from the establishment of his sovereignty."[38] This was again a long and arduous process.

A different type of mastery over the natural world was deduced from prehistoric artworks. "Mobile art" of carved bones and etched ivory had been discovered alongside the establishment of human antiquity itself. These records were expanded throughout the latter part of the nineteenth century and were even more strikingly demonstrated during the slow recognition of the authenticity of cave-art sites like Altamira and Font-de-Gaume from the 1880s to the 1900s.[39] This was important on a number of levels for documenting human interaction with ancient animals. First, it gave the most direct proof possible of contemporaneity and also showed that many creatures were of great interest for prehistoric "man" (artists were almost always assumed to be male). It was also often tied with a notion of conceptual mastery. While some early interpretations of prehistoric art saw it as the product of "idleness," representing the thoughtless activity of bored savages between hunts, in the latter part of the nineteenth century it was often accorded a religious or ceremonial significance, with the French prehistorian Salomon Reinach (1858–1932), interpreting cave art as fetishistic "hunting magic." Similar ideas were presented in Britain, where scholars such as William Johnson Sollas (1849–1936) gave a great deal of attention to the possible ritual significance of Paleolithic artworks. In addition to placing the origins of religion and spirituality in these deep periods, the recognition of Paleolithic art led to a great deal of admiration for the hunter-artists of the Old Stone Age. One work of 1912, H. G. Spearing's *The Childhood of Art*, was to state that "in ancient times the prim-

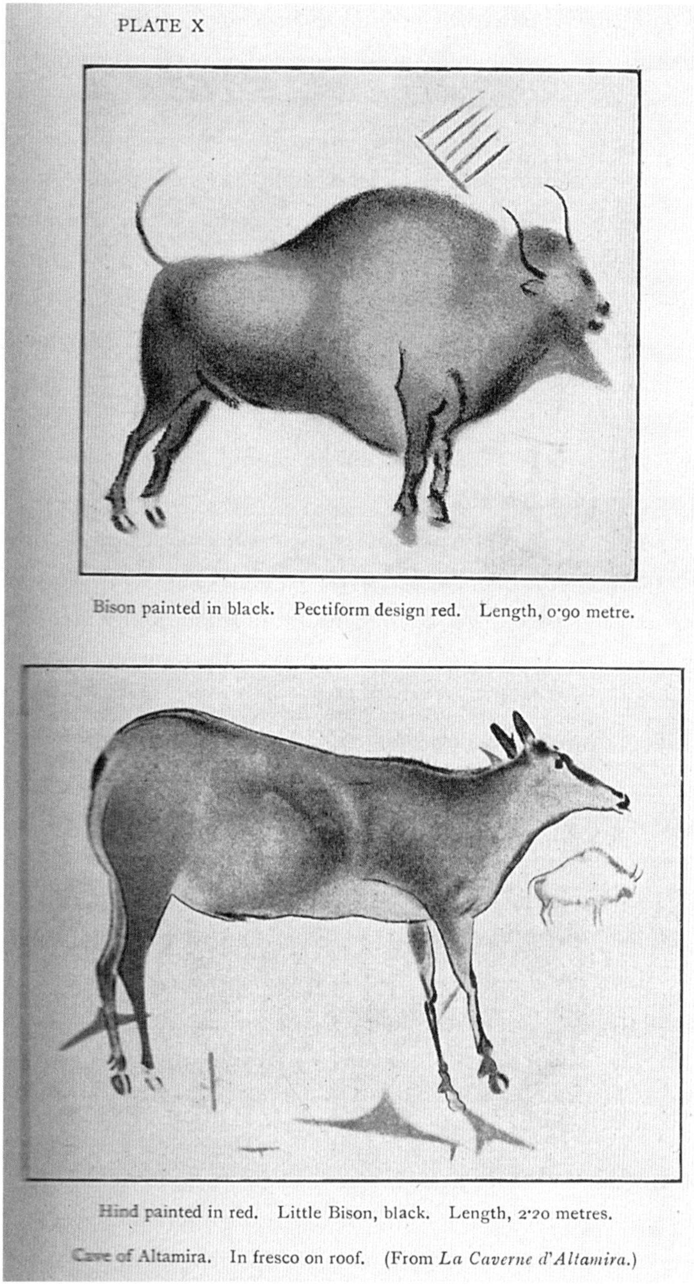

FIGURE 1.6. Image 1 from the Cave of Altamira. From Ernest A. Parkyn, *An Introduction to the Study of Prehistoric Art* (London: Longmans, Green, 1915), 98–99. Author's collection.

98 PREHISTORIC ART

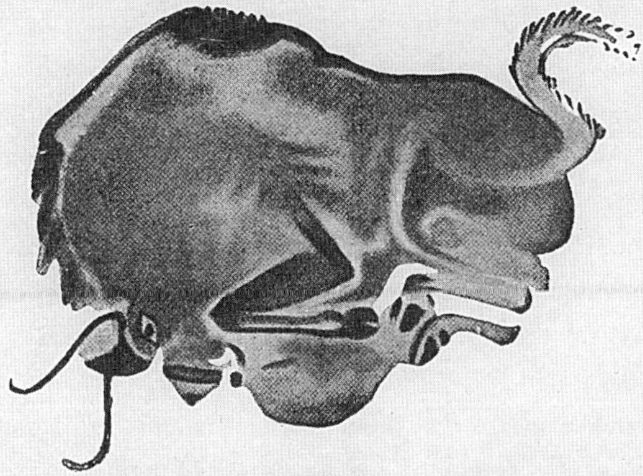

Fig. 127.—*Altamira*. Bison jumping, polychrome.

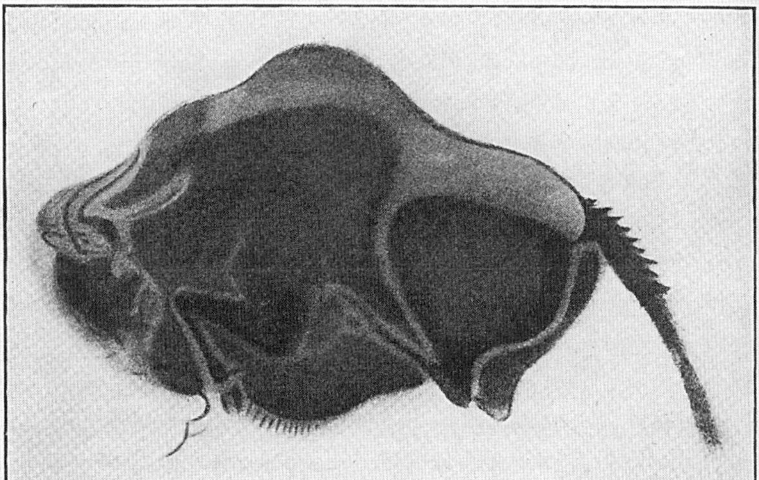

Fig. 128.—*Altamira*. Bison in polychrome. Great fresco. (From *Cav. d'Altam.*)

FIGURE 1.7. Image 2 from the Cave of Altamira. From Ernest A. Parkyn, *An Introduction to the Study of Prehistoric Art* (London: Longmans, Green, 1915), 98–99. Author's collection.

itive artist was evidently a mighty hunter as well as a close observer of wild animals. He had watched the mammoth trampling through the forest and he had seen the bison stand at bay. He had faced the wild boar's frenzied rush and the onslaught of the wounded stag. And what he saw he remembered, noting the curve of the back and the poise of the head, the firm planting of the massive hoof, or the twinkling motion of the legs of graceful deer."[40] Skill as a hunter and an artist operated in synergy, and both drew from interaction with prehistoric animals.

Paleolithic art was also significant in another respect. Given its highly realist nature, it provided additional evidence of the extinct animals "in life," showing their physical features. A particularly famous example was the mammoth carved in ivory found at Les Eyzies in the Dordogne in 1864. As well as giving incontrovertible evidence that humans had seen and lived alongside prehistoric animals, it also developed understandings of mammoths more broadly, through giving detailed observations of their physical characteristics. Dawkins was to write of this object:

> The most striking figure that has been discovered is that of the mammoth, engraved on a fragment of its own tusk, the peculiar spiral curvature of the tusk and the lone mane, which are not now to be found in any living elephant, proving the original was familiar to the eye of the artist. The discovery of whole carcasses of the animal in northern Siberia, preserved from decay in the frozen cliffs and morasses, has made us acquainted with the existence of the long hairy mane. Had not it thus been handed down to our eyes, we should probably have treated this most accurate drawing as a mere artist's freak. Its peculiarities are so faithfully depicted that it is quite impossible for the animal to be confounded with either of the two living species.[41]

In this respect, objects created by ancient humans became important evidence in the paleontology of the Pleistocene, showing details that were impossible to gain from simply studying bones. Evidence such as this was increasingly brought to visual reproductions of prehistoric animals and ensured that Paleolithic artworks not only documented human-animal interaction but served to increase empathy with the peoples of prehistory—almost bringing them on as collaborators in paleontological reconstruction.

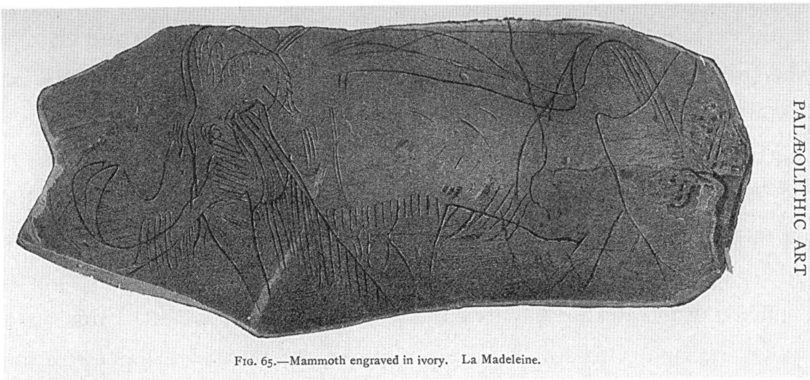

FIGURE 1.8. "Mammoth engraved in Ivory. La Madeleine." From Ernest A. Parkyn, *An Introduction to the Study of Prehistoric Art* (London: Longmans, Green, 1915), 50. Author's collection

THE FATE OF THE ANIMALS

Understandings of extinct beasts had been essential for establishing the idea of human antiquity and conceptualizing how prehistoric humans lived. However, looming over these discussions was another issue: What exactly had happened to all of these creatures, and why had they died out? The strangeness of the Pleistocene fauna, mixing together its "Arctic" and "African" animals, was in stark contrast to that of modern Europe (and even the entirety of the Northern Hemisphere). Some of the animals characteristic of the Pleistocene, such as the mammoth and the woolly rhinoceros, had died out completely, while others, such as the reindeer, musk ox, hippopotamus, and hyena, had apparently receded northward or southward. Whether through extinction or migration, the image was of steady decline in large animals over recent geological history. Indeed, the paleontological chronology developed by Édouard Lartet was predicated on this idea, based on the initial presence and then subsequent disappearance of the cave bear, the reindeer, and the aurochs. More detailed studies showed an even more pronounced line of extinctions, with mammoths, hippo, woolly rhinoceros, cave bear, hyena, lion, Megaloceros, and reindeer all sequentially vanishing over prehistory. The extinctions did not stop then but continued into historic periods, with the aurochs, bison, beaver, wolf, and brown bear also being eliminated in much of Europe in the centuries prior to the nineteenth. The chronologies of prehistory and history, when placed together, showed a steady decline in animal life, which continued into the present.

Much of this was understood in terms of an epochal transformation, as the "Age of Mammals" gave way to the "Age of Man." The decline and disappearance of the large beasts and ferocious carnivores were necessary for the growth of human society and civilization, with the wildness and diversity of the fauna decreasing as humans gained "lordship" over the Earth. This was a notion similar to Richard Owen's concepts of providential change, with the next stage of Earth's development—its use by "man"—requiring a reordering in the dominant forms of life. All of the models for human interaction with prehistoric animals discussed above presented the Pleistocene fauna as threatening for human social organization and the growth of civilization. It was therefore necessary for the animals to disappear in order for human society to reach its full potential.

However, despite this, there was a possibly surprising reluctance to assign humans full responsibility for the extinctions. It appears as if there was some uneasiness over both the moral responsibility for extinction and doubts over the capacity of primitive humans to fully wipe these animals out. This often made commentators turn to other mechanisms. Sometimes the line of extinctions was stretched even further back—to the highpoint of diversity in the Miocene (a period some way before the Pleistocene and before most chronologies placed human origins). This meant that the continuum of decline was much longer, had initially occurred without human impact, and was therefore due to other forces. Climatic change was frequently presented as an important mechanism, either making the environment unsuitable for warm-adapted animals or altering the amount and type of vegetation available as food for large herbivores. Disease was also sometimes brought in as an additional cause to explain the decline of particular animals. Finally, metaphysical ideas of providential decline were invoked to explain why particular lineages or faunas had died out. Many of the early commentaries were highly providential, with some divine figure or scheme making the Earth ready for humans and disposing of the larger, more threatening animals. In the 1890s and 1900s scientists also began to argue for the notion of "evolutionary senility," that organisms reached an evolutionary stage where they could adapt no further and could only decline and quietly die as conditions changed. These large processes driving extinction were invoked frequently, although which had precedence was again far from agreed upon.

Humans were not totally excluded from these extinctions, however, and many discussions did have early humans accelerating something, which was already beginning. As early as 1860, Richard Owen modified his views along these lines, taking in the establishment of human antiquity to explain the decline of certain large animals. He noted: "As a cause of extinction in times anterior to man, it is most reasonable to assign the chief weight to those gradual changes in the conditions affecting a due supply of sustenance to animals in a state of nature which must have accompanied the slow alterations of land and sea brought about in the aeons of geological times."[42] This was primarily due to long-term geological changes, which worked in a Malthusian manner affecting the amount of food available for large animals. However, Owen then continued: "Recent discoveries indicate that, in the case of the last two extinct quadrupeds [mammoths and woolly rhinos], a rude primitive human race may have finished the work of extermination, begun by antecedent and more general causes."[43] Prehistoric hunters were therefore potentially responsible for eliminating the last of the larger herbivores. However, this was still a complex multicausal process, which required a long geological backstory prior to the appearance of humans and the final extermination of these creatures.

Similar arguments were made by John Lubbock in *Pre-Historic Times*, although he made greater allowances for the differing effects of human hunting and environmental change on different types of creature. He noted how, since the Pleistocene, "our climate has greatly changed for the better, and with it the fauna has materially altered. In some cases, for instance, in that of the hippopotamus and of the African elephant, we may probably look to the diminution of food and the presence of man as the main causes of their disappearance; the extinction of the mammoth, the *Elephas antiquus*, and the *Rhinoceros tichorhinus*, may possibly be due to the same influences; but the retreat of the reindeer and the musk ox are probably in great measure owing to the change of climate."[44] Again, the diversity of the fauna ensured that different mechanisms would be relevant for different creatures. Larger animals were felt to be more vulnerable to changes in the climate, as they required a great deal of food to sustain themselves. They were to be the first to die when major shifts occurred in the environment. Extinction through hunting, however, was difficult to invoke with regard to the large herds of reindeer and musk-ox upon which Ice Age humans were so dependent. That these animals were also so clearly adapted for cold climates—and still lived in northern regions in interaction with modern humans (many of whom, such as the Sami, were judged by comparative analogies as being similar to the inhabitants of Paleolithic Europe)—also worked against ideas of invariable extinction. As human society and the environment in the modern Arctic North were understood as closely aligned with that of "reindeer age" Europe, the relationship between climate, humans, and the animals was stable and could not result in extinction.

Whatever the causes of early extinctions, human responsibility for the destruction of animal life increased alongside the growth of civilization. Here the interaction was twofold. On the one hand, the clearing out of wild spaces for agriculture and settlement was seen as diminishing the ability of animal life to survive and multiply. However, other animals were domesticated to become accessories and partners with human society, and human reliance on wild animals became replaced by management of domesticated creatures. Paleontological analyses of later archaeological periods such as the Neolithic and the Bronze Age presented the iconic animals found alongside humans as not being wild animals but those in the early stages of domestication, particularly dogs, cattle, and horses. Charles Lyell (1797–1875), reporting on the studies undertaken by Ludwig Rütimeyer (1825–1895) in the 1850s and 1860s of the Swiss Lake-Dwellings, noted this transformation, that in the sites from "the earliest age of stone, when the habits of the hunter state predominated over those of the pastoral, venison, or the flesh of the stag and roe, was more eaten than the flesh of the domestic cattle and sheep. This was afterwards reversed in the later stone period and in the age of bronze."[45] Analyses of animals over a long period indicated the increase of civilization and settlement, moving from the "hunting state" to that of the pastoralist of agriculturalist (a key transition in stadial models of development, which still strongly impinged on ideas of prehistory).

While a full investigation of Victorian understandings of the history of animal domestication is outside the scope of this chapter, this change does indicate that once agriculture and settlement had developed, human and animal interactions began to be conceptualized in a different manner—with wild animals becoming separate from humanity, to be replaced by domesticated ones "tamed by man, and [who] served him either as auxiliaries or companions."[46] For Louis Figuier, this built into a triumphant account. His work concluded with the settled agricultural society of the Iron Age, and ended:

> And now, O man, thy work is nearly done! The mighty conflicts against nature are consummated, and thy universal empire is for ever sure! Animals are subject to thy will and even to thy fancies. . . . Thou hast changed the whole aspect of the globe, and mayst well call thyself the lord of creation!
>
> Doubtless the expanding circle of thy peaceful conquests will not stop here, and who can tell how far thy sway may extend? Onward then! still onward! proud and unfettered in thy vigilant and active course towards new and unknown destinies!
>
> But look to it, lest thy pride lead thee to forget thy origin. However great may be thy moral grandeur, and however complete thy empire over a docile nature, confess and acknowledge every hour the Almighty Power of the great Creator. Submit thyself before thy Lord and Master, the God of goodness and of love, the Author of thy existence, who has reserved for thee still higher destinies in another life. Learn to show thyself worthy of the supreme blessing—the happy immortality which awaits thee in a world above, if thou hast merited it by a worship conceived in spirit and in truth, and by the fulfilment of thy duty both towards God and towards thy neighbour!![47]

The growth of human civilization, coupled with the extinction of ferocious wild animals and the domestication of tamable ones, was connected with economic development, prosperity, and the furtherance of divinely ordained providence. As with Dawson, this was also interpreted in a scripturally informed manner and shows the ways in which a theologically defined conception could be maintained within a framework that emphasized the material aspects of human progress over nature.

However, these ideas that only animals useful to humans would survive as the "age of man" progressed were not just presented in triumphalist manners. Particularly as the nineteenth century moved on, and doubts around the potentials of human progress began to manifest strongly, prehistoric extinctions were discussed in a more ambivalent or melancholy manner—particu-

FIGURE 1.9. "Cultivating Gardens during the Bronze Age." From Louis Figuier, *L'homme primitif* (Paris: Librairie de L. Hachette, 1870), 329. Author's collection.

larly when understood in terms of contemporary extinctions or the decimations of animal life in the Americas and Africa. This was very much the case in Harold Knipe's *Evolution in the Past*, which concluded with a gloomy prognosis for the future of "brute life," which had gone through a major decline across the prehistoric ages to the point where "to-day brute-life, except in a domesticated condition, is at a very low ebb." He acknowledged that "this long-continued downward course is doubtless in part ascribable to climatic changes and disease; but," he continued, "in its later stages it has beyond question been brought about chiefly by man." A continued comparison followed across all the different animal species:

> How great the fall has been may also be gathered from the present distribution and condition of some familiar animals. Elephants, at one time consisting of many species, were prosperous in Europe, Asia, Africa, and North and South America. Now they are reduced to one African and one Indian species. As the family has shown itself willing enough and able to co-operate with man, the

> majestic old line may long continue. Rhinoceroses, once ranging freely over several continents, are now known only in restricted areas of Africa, India, and the Malayan countries. Their range, too, is bound to diminish, for the animals have no stomach for civilization, and are never likely to be seen drawing the plough, or otherwise toiling in man's service. . . .
>
> Man meanwhile has made a wonderful progress. Master of brute-life, he has also acquired no small control over the blind forces that pervade the universe; and in various other ways he has travelled far from a purely animal condition. It cannot, however, yet be said that he has gained complete mastery of his brute passions and impulses which he has inherited from a remote past. The achievement of this, so far as can be seen, is an immediate purpose in his further evolution.[48]

This did not deny that the end of the large animals was a corollary of human progress and possibly "necessary" for further developments in wider cosmic evolution. However, it gave a tragic tone to these processes, as majestic, large, and formerly extensive creatures receded and finally disappeared. Conservation could only be a temporary expedient, and only tamed animals would be able to survive in the long term.

In some respects, this account of nineteenth-century understandings of human interaction with prehistoric animals shows a familiar story of the nineteenth-century faith in progress and the idea that human existence not only took precedence over the animal but represented a providential transformation. The evidence of human contemporaneousness with prehistoric animals illustrated a steady decline in the size and ferocity of animal life, which was almost correlated with the increasing sophistication of human tools and social organization. Animals were felt to be declining over recent geological eras, and human ascendancy over the natural world was linked with this, being the next step in a progressive trajectory—whether this be ordained through overtly religious, vague providential, or evolutionary principles. In this respect, concepts derived from naturalistic scientific ideas of change in the natural world and more theological notions of human providential mastery could both be expressed in a language of development and transition. Humans were constantly presented as at least aspirational masters of the world, and the slow developments over prehistory showed an increased mastery over nature as human civilization, culture, and technology progressed. This progress of human society was quite consistently presented as directly opposed to wild nature, and the growth of civilization was contingent upon the extinction of large and dangerous animals, the clearing of wild spaces, and the use of medium-sized mammals either for hunting or as domesticates.

However, there were also other notions, which have been less well appreciated in much of the secondary literature. First, the spectacular nature of the animals of prehistory and the sensation around prehistoric discoveries meant that the interaction of prehistoric humans and animals was often presented in a dramatic idiom. Particular periods of human prehistory were felt to be directly connected with particular forms of animal, whether these be the "men among the mammoths," the reindeer hunters, or the first domesticators. This locked the stages and development of European prehistory in with particular creatures and showed "savage" humans to be dependent on and defined by their relations with animals, through both the development of chronologies and the current understandings of their lifestyles. This also ensured that wonder and artistic engagement with prehistory—whether through imaginative reconstructions of the entire fauna, visual representations of interactions with particular types of prehistoric creature, or consideration of the artworks produced by the prehistoric peoples themselves—locked the inhabitants of Ice Age Europe in relationships with a range of animals. While increasingly dominant over the natural world, early "savage" humans were also a definite part of it, and the more snapshot views of particular eras showed humans as often being a part of the fauna rather than "masters of creation."

Second, and possibly more telling, much of this discourse also showed a marked trepidation—or uncertainty over human capacities. In early phases of prehistoric development, as discussed by Worthington Smith and some sections of Figuier, humans were often presented as having been incapable of acting against the fierce nature of the Pleistocene world and in need of environmental change, evolutionary development, and improved technology to finally give them the edge. When combined with the varied discourses around animal extinction, this often meant that the growth of civilization had to be thought of as being as much dependent upon providential and climatic factors as upon any intrinsic human capacities. Furthermore, comparing the deep evolutionary history of mammals and the rich fauna of prehistoric Europe with the diminished animals of the modern world was often a bleak prospect. Particularly later in the period, there was a strong sense of loss and tragedy around the decline of the great animals, which had been such an important feature of life for humanity's early ancestors but had now vanished or been pushed to the margins of the world. While this was seen as necessary for "progress" and the onset of history, it nevertheless meant that a key part of the former world was lost.

2

OF ROCKS AND "MEN"

The Cosmogony of John William Dawson

NANNA KATRINE LÜDERS KAALUND

In 1860 the Canadian geologist, paleontologist, and university administrator John William Dawson (1820–1899) published what would be the first of many substantial publications he produced as a popularizer of science, *Archaia; or, Studies of the Cosmogony and Natural History of the Hebrew Scriptures*. The central theme in *Archaia* was to show that on the question of human origins the Bible and science, when interpreted correctly, were in agreement and that all humans originated from a single pair of ancestors, Adam and Eve. Dawson's historicization of humans sought to discredit polygenism, which asserted the plural origins of humans, while simultaneously showing that when materialism, polygenism, and polytheism were discarded, one can observe a fundamental harmony between Scripture and science. *Archaia* was a commercial failure, but it is a significant example of the way one of the most influential nineteenth-century scientific figures in Canada viewed the nature and content of scientific practice.[1] Dawson had a particular vision for science in Canada. Using his leading position within the Canadian scientific community as principal of McGill University in Montreal, he aimed to direct those interested in the study of human origins, history, and religious studies by enrolling them into his theoretical model. The aim here is to look at how Dawson discussed human origins and history in *Archaia*, and to explore his double commitment to science and the Bible. What emerges is a revealing story about the relationship between science and religion in the mid-nineteenth century that gives insight into the type of strategies utilized to reconcile science and Christianity, before and after the publication of Charles Darwin's (1809–1882) *On the Origin of Species* in 1859.

Studies on antitransformist writers in the nineteenth century often center on their relation to Darwin's *Origin of Species*. However, his theory was not the only challenge to more literal readings of the Genesis narrative, and Dawson's position did not become untenable after 1859. Although *Archaia* was published the year after *Origin of Species,* it did not include references to Darwin's theory aside from a footnote focusing on Darwin and Alfred Russel Wallace's joint paper in the *Linnean Transactions* from 1858. *Archaia* did, however, still directly address the question of developmentalism and transmutationism, with references to such works as James Cowles Prichard's *Researches into the Physical History of Man* (1813), and Robert Chambers's anonymously published *Vestiges of the Natural History of Creation* (1844). While *Archaia* had been prepared already in 1855, it is suggestive that Dawson did not add a discussion on *Origin of Species.* As James Secord has shown in *Victorian Sensation,* the tendency to have the history of evolution begin with Darwin—or to see the *Origin of Species* as the book that solely made evolution well known and accepted—is misguided. Similarly, in *The Earth on Show,* Ralph O'Connor has emphasized the pitfalls of looking only to (or from) Darwin. In doing so, O'Connor shows the significance of lesser-known figures such as the geologist and editor Hugh Miller (1802–1856), who was a great influence on Dawson. In 1860 Dawson famously reviewed *Origin of Species* in the *Canadian Naturalist and Geologist,* and the updated version of *Archaia*—which was published in 1877 as *Origin of the World According to Revelation and Science*—actively engaged with Darwin's work and that of the scientific naturalists Thomas Huxley (1825–1895) and John Tyndall (1820–1893).[2] Yet Dawson's *Archaia* shows the distinct challenges facing a literal or creationist reading of the Genesis narrative, outside of the influence of *Origin of Species.*

This chapter, therefore, is connected to the broad historiography on the relationship between science and religion in the nineteenth century. It further builds upon Susan Sheets-Pyenson's important biography of Dawson, and the work of Suzanne Zeller and Charles F. O'Brien, who have shown the significance of Dawson's research, his theological position, and his impact on Canadian science and society.[3] Dawson's process of historicization was linked to these themes; and *Archaia* was significant, in spite of its poor commercial success, as an expression of how Dawson's views on human history were radically different from what was happening in the scientific milieus in Britain. The conflict thesis between science and religion—epitomized by John William Draper's *A History of the Conflict between Religion and Science* (1874) and Andrew Dickenson White's *The Warfare of Science with Theology in Christendom* (1896)—has been the subject of extensive scholarly research. As the conflict thesis has been debunked as a valid historical interpretation of the relationship between science and religion, scholars such as Frank Turner, Bernard Lightman, Peter Harrison, Crosbie Smith, Jonathan Topham, Geoffrey Cantor, and John Hedley Brooke have shifted focus away from the narrative of conflict to the range of epistemological redefinitions of what was considered to constitute proper scientific practice and the scientific community in the nineteenth century.[4] Figures such as Huxley sought to push religion out from scientific practice in Britain. But these attempts were not wholly successful. Moreover, as Lightman, Aileen Fyfe, and others have shown, many popularizers of science promoted a view of nature that was explicitly theological.[5] Examining Dawson's *Archaia* in the context of Canadian and British science shows the significance of religious ideas in the historicization of humans. The pervasiveness of literal readings of the Bible in Dawson's work (and its popularity) problematizes the narrative of scientific secularization and opens up questions about the state of science in Canada during the nineteenth century.

Dawson argued that geology needed the Bible and that theology required geology. He considered the Deluge a historical fact, he was an antitransformist throughout his life, and he believed humans had been on the Earth for only a brief period of time—although he allowed for a longer prehuman history of the Earth. The ways nineteenth-century writers tackled the issues of the time frame of the Earth and the place of humans in this time frame have been the subject of a large secondary literature, including recent interpretations of historians Ralph O'Connor, James Secord, Ronald Numbers, and David Livingstone.[6] Dawson positioned himself as an authority on the question of human history, particularly because of his knowledge of both Scripture and science. He considered the two major issues involved in reconciling the cosmogony of the Bible with the teachings of science to be the early history of the Earth and the history and nature of "man." While answering these two questions, *Archaia* simultaneously outlined a framework for the nature of proper scientific practice and added Dawson's vision of science to the body of scientific knowledge in Canada. The naturalists who refused to acknowledge the importance of the Bible for the advancement of science were just as bad as those theologians who considered the study of nature to be damaging for faith and who held geology as one of the most problematic sciences for Christianity. For Dawson, good scientific practice was that which found an answer to the fundamental question of what a human is in the careful study of nature and the Bible, and this was what he sought to advance in Canada.

A VISION OF CANADIAN SCIENCE

> *Now for the first time was the earth tenanted by a being capable of comprehending the purposes and plans of Jehovah, of regarding his works with intelligent admiration, and of shadowing forth the excellences of his moral nature.*[7]

Dawson was born in Nova Scotia in 1820 to Scottish parents of strong Presbyterian faith. He attended the University of Edinburgh where he studied natural history, and after graduating in 1842 he returned to Nova Scotia. Within the Canadian context, Dawson was a pioneer of geological exploration and an influential naturalist and university administrator.[8] Dawson took a keen interest in advancing education and science in Canada, and he served as the principal of McGill University in Montreal between 1855 and 1893. He was a member of several learned societies, including the Geological Society of London, where he was elected fellow in 1854. He was also elected as president for multiple scientific meetings and societies, such as the Natural History Society of Montreal, the American Association for the Advancement of Science, and the British Association for the Advancement of Science. In 1881 he was awarded the Lyell Medal by the Geological Society of London, and he was the founding president of the Royal Society of Canada (established in 1882). Dawson was intent on advancing science in Canada, a task that included establishing scientific societies, developing educational administration to cover more scientific topics, and publishing scientific works in Canada through Canadian publishers.

In the preface to *Archaia*, Dawson wrote: "The work is issued in Canada, because the writer desires to contribute his mite to the growing literature of British America, and has found in Montreal a house sufficiently enterprising to undertake the risk of publication."[9] Using a Canadian publisher was a key part of Dawson's strategy for advancing his vision of Canadian science. As with the British context, a range of factors influenced the growth of scientific and general publishing in nineteenth-century Canada, including changes in print technologies, rapid transatlantic and railway services, and increased literacy.[10] The context for science and scientific publishing in Canada in the nineteenth century was shaped by the political turmoil of that period. While science was a popular topic in the periodical press in Britain, both in specialized journals and general newspapers, this was not the case in Canada.[11] Although there were hundreds of specialized periodicals in print in Canada in the second half of the nineteenth century, only a few of these were dedicated to scientific topics. The editors of the collection *The History of the Book in Canada* have suggested that this was because it was seen as more prestigious to publish in the American and British journals.[12] The publisher of *Archaia* was the Dawson Brothers (no relation to John William). Susan Sheets-Pyenson has emphasized the central role of the Dawson Brothers, as "the publication of *Archaia* also initiated Dawson's long and close relationship with the publishing firm . . . who became known for their scholarly and scientific editions."[13] The Dawson Brothers was also the publishing house behind the important Canadian periodical publication, the *Canadian Naturalist and Geologist* (est. 1856).[14] As part of the editorial committee of the *Canadian Naturalist and Geologist*, the journal and the Dawson Brothers were central players in Dawson's quest for directing Canadian science. The significance of *Archaia* being published in Canada was not lost on the reviewers, and one writer for the *Athenaeum* noted that its publication "speaks well for Science in Canada."[15]

Archaia was published in the popular octavo format (crown 8vo), bound in cloth, and cost seven shillings and sixpence. A cheaper edition at six shillings was published the same year by Samson Low in London. Covering 397 pages including an extensive appendix, it was structured in more than 17 chapters that took the reader through an account of the history of the world, from each day in the Genesis narrative through a discussion on the nature of "man." Aside from the introduction, each chapter began with a quote that set the tone for the discussion. The method Dawson applied to the study of the Bible and science is easily exemplified by the opening quote to chapter 2, "Objects, Character and Authority of the Hebrew Cosmogony," by Sir T. Browne: "There are two books from which I collect my divinity; besides that written one of God, another of his servant nature—that universal and public manuscript that lies expansed unto the eyes of all."[16] Dawson's tactic for making geology and the study of humans a safe topic by showing how the sciences, when interpreted his way, did not significantly clash with the Bible was not unique. In *Science and Salvation* (2004) Aileen Fyfe has shown the significance of the "two book" metaphor for the relationship between science and faith outside of the community of specialists. In doing so, Fyfe demonstrates that by the mid-nineteenth century, the evangelical theological framework included the sciences. The worry, Fyfe argues, was not so much about the specific scientific discoveries but, rather, about how these were presented to a wider reading audience to avoid constituting a danger to salvation.[17] Similarly, Dawson believed that careful studies of science and the Bible would be mutually beneficial, not only for the advancement of knowledge but also for the moral status of humans. The study of the Bible was as necessary for the salvation of scientific men as the study of nature was for students of divinity.

Curiously, Dawson argued that his book was not an attempt at reconciling science and the Bible. In the introduction, he wrote that the character of his book was such that it examined the facts relating to humans and Earth history, as they appeared in natural science and the Bible. However, it was not an attempt to establish a scheme of reconciliation between science and Scripture. There was no need to bend or modify the findings of natural science to fit with the Bible, as both the findings and the Bible were in their way a true record of the history of the Earth. Furthermore, Dawson argued that naturalists should remember their obligations to the Bible and should prefer the Bible to modern pantheism, positivism, and priestly authority.[18] In those instances where a discord appeared, he noted, it was enough to simply reflect upon its existence and assume that future discoveries would reveal those instances to be in accordance after all.[19] Chapters 4 through 16 began with a verse from Genesis that corresponded with the subject examined in the chapter, while chapter 17, "Comparisons and Conclusions," ended with "Lo these are but outlines of His ways, and how faint the whisper which we hear of Him—the thunder of His power who could understand" (Job 26:14).[20] The implication was that *Archaia* had outlined the work of the Creator in a way that allowed the reader to understand the double testimony from nature and Scripture.

Dawson was a leading scientific and administrative figure in a period often thought of as a time when elite science in Britain was removed from Christianity. Fyfe and Lightman have shown the pervasiveness of arguments for design in scientific books, especially among the popularizers of sciences in late nineteenth-century Britain, but Dawson was writing from within the elite scientific establishment in Canada.[21] This is significant, because Dawson's *Archaia* shows the difference between the scientific milieus in Canada and Britain. Dawson's position was similar to that of the Swiss American geologist Arnold Guyot (1807–1884) at Princeton and the American geologist James Dwight Dana (1813–1895) at Yale. When viewed within the broader North American context, Dawson was not as much of an outlier. Yet, as Numbers has argued, Guyot and Dawson were "the last of the nineteenth-century scientists to cling to creation."[22]

Within the larger British world the picture was different. What was happening in Britain was not necessarily representative of what was happening in its empire. The picture shows the complicated nature of the relationship between science and religion, and between the elite and the popular, in the nineteenth century. Suzanne Zeller has emphasized the relationship between science and nation building in Canada, arguing that "science became the gauge by which Canadians assessed what their country, and, through it, they themselves could one day become."[23] Shaping Canadian science, before confederation, was thus a key aspect of shaping Canada. Dawson, from a position of authority in Canada, could write an account of human history that was very different from those produced by his British contemporaries within the elite scientific establishment. *Archaia* was not an exercise in reconciliation, he argued, as there was nothing to reconcile. It was this version of science, one that drew from and added to biblical exegesis, that Dawson wished to advance in Canada.

> *In short, we may take this as a plain and authoritative declaration* that the day of creation is not the day of popular speech.[24]

Dawson drew parallels between the days of the Bible and geological periods of Earth history, outlining what is termed a "day-age theory." In his exposition of the history of the Earth, a key element was to establish a time frame for creation that was in accordance with both Scripture and geological records. Dawson was a proponent of what is termed the day-age system; when interpreting the meaning of "days" in Genesis he drew parallels between the days of the Bible and the geological periods. Three key figures influenced Dawson's day-age theory: Hugh Miller, Charles Lyell (1797–1875), and James Hutton (1726–1797). By linking the days of Genesis to geological periods, Dawson sought to preserve the integrity of the creation narrative. Moreover, by placing the creation of humans late on the sixth day of Genesis, Dawson was able to argue that geology and the Bible were in agreement, and that humans had only been on the Earth for a short period of time. This was a key point for his process of historicizing humans.

Together with Dana, Dawson was a key figure in popularizing the day-age or day-epoch interpretation of the Earth's past in North America in the second half of the nineteenth century.[25] This was not a secular version of human and Earth history but was openly constructed with the biblical constituting at least an equal part to the evidence from science. The days in Genesis were not like the days of common speech; rather, they were long and undefined periods of time.[26] The view that the days in Genesis should be considered periods or epochs has a long history, reaching back to St. Augustine (354–430).[27] *Archaia* did not contain a detailed account of the previous works on day-age theories, and Dawson stated instead that he would "prefer to direct [his] attention immediately to the record itself."[28]

One of the key figures that Dawson did explicitly discuss was Hugh Miller.[29] O'Connor has shown in *The Earth on Show* how Miller developed several literary techniques for geological popularizers, including particular ways of manipulating the "reader's visual imagination" by drawing on the popular panoramas and dioramas, thereby linking theater and storytelling.[30] Dawson made several references to this way of writing in his discussion concerning the meaning of "days" in Genesis. For example: "On the theory of days of vision, as expounded by Hugh Miller, in the Testimony of the Rocks, in one of his noblest passages, the evening and night fall on each picture presented to the seer like the curtain of a stage."[31] Dawson quoted long passages from Miller, and their views of the relationship between science and religion were similar. As with Dawson, Miller considered geology and Scripture to be allegories of each other, both leading the reader to Christ. Modern geology was the key to unlocking and enriching (not eclipsing) Genesis.

According to Dawson, aside from geology there were two sciences, "which have in modern times attempted to penetrate into the mysteries of the primitive abyss," and these were astronomy and chemistry.[32] However, none of those sciences could be anything but hypothetical about the early state of the Earth. Dawson held that the laws of nature first came into existence on the fourth day of creation. Since their first creation, the laws of nature had remained constant. On this matter, Dawson argued, the Bible and geology were in accordance. In the Bible the laws of nature were continuous and invariable, representing an unchanging creator. Similarly, modern geology shows that there was a "higher law of progress." Dawson's view that "during the whole time referred to by geology, the great laws both of inorganic and organic nature have been the same as at present" was similar to what Charles Lyell outlined in *Principles of Geology* (1833) as a necessary principle for making geology a true science: uniformity as a philosophical foundation of geology.[33] Lyell argued that the study of the Earth had to be done under the assumption that causes now visible around us are the same as those that acted in the past, and that they act with the same intensity now as they did then. The key issue was "causes," which was in line with Isaac Newton's notion of "true causes" or *vera causa*.[34] James Secord has noted that although Lyell underlined the immensity of past time, he did not view the Earth as eternal, and neither did he present *Principles of Geology* as a cosmological treatise but, rather, as a manifesto for what pursuing science should be like.[35]

Dawson significantly diverged from Lyell on three points. First, while Lyell argued that there was no need to use intermittent catastrophes as explanations for changes in the Earth's appearance, Dawson firmly held that the Deluge was a historical reality. Second, for Lyell, it was imperative to distance himself both from Jean-Baptiste Lamarck (1744–1829) and from Scripture, but for Dawson Scripture should be equal to the analysis of nature in the science of the Earth. As Secord has shown, the intended audience of Lyell's *Principles of Geology* was a conservative, genteel, and professional audience, and "this audience needed to be shown that geology was safe."[36] This meant explicitly distancing himself from Lamarck. With regard to Scripture, Lyell believed that one way of making science respectable was to make it philosophical, and he considered the clergy to be the chief obstacle in this pursuit.[37] While Lyell did not argue for an inherent conflict between science and religion, his position was that of separate spheres.[38] Third, while Dawson also focused on establishing what he thought was the best form of geology, *Archaia* extended the account of the history of the Earth back to the earliest time of creation. While Lyell and Dawson diverged on such points, they maintained a friendship, collaborating on geological explorations, and exchanging letters. Thus, for Dawson, Lyell was an important ally in England.[39]

While Dawson argued that evidence from the Bible and science should be equal, there were certain aspects of the study of the Earth's history that could only become known from interpretation of Scripture, namely, what the world was like during its earliest days, and where its future was heading. Dawson held that science could say nothing of the state of the Earth prior to the creation of the laws of nature. Geology, Dawson argued, could account for the history of the Earth only after the introduction of the dominion of existing causes on the fourth day of creation:

> Geology, as a science of observation and induction, does not carry us back to this period. It must still and always say, with Hutton, that it can find "no trace of a beginning, no prospect of an end,"—not because there has been no beginning or will be no end, but because the facts which it collects extend neither to the one nor the other. Geology, like every other department of natural history, can but investigate the facts which are open to observation, and reason on these in accordance with the known laws and arrangements of existing nature. It finds these laws to hold for the oldest period to which the rocky archives of the earth extend. Respecting the origin of these general laws and arrangements, or the condition of the earth before they originated, it knows nothing.[40]

As with Lyell, Dawson argued for the uniformity of the laws of nature, but Dawson also argued that there was a long period where these laws of nature did not exist, and hence science could not examine the state of the world at this time. It was only since the period corresponding to the fourth day of Genesis that the laws of nature had remained constant, which allowed science to trace the history of the Earth.[41] This position restricted the scope of geology, and any other science for that matter. Only the Bible could provide clues as to the very first origins of the Earth.

Another key influence on Dawson's geology was James Hutton. Mott Greene has described Hutton's theory of the Earth as "a classic of deistic natural theology and natural philosophy in which geological phenomena were evidence for the conclusion that no system so balanced and purposeful could ever have emerged without intelligent design."[42] As with Hutton, Dawson saw in nature proof of its intelligent design, and especially in geology:

> In the department of "final causes," as they have been termed, scripture and geology unite in affording large and interesting views. They illustrate the procedure of the All-wise Creator, during a long succession of ages, and thus enable us to see the effects of any of his laws, not only at one time, but in far distant periods. To reject the consideration of this peculiarity of geological science, would be the extremest folly, and would involve at once a misinterpretation of the geologic record, and a denial of the agency of an intelligent Designer as revealed in scripture, and indicated by the succession of beings.[43]

Dawson's day-age system began with the creation of matter. The second day of Genesis accounts for the formation of the atmosphere, at a point where the Earth was covered by water, which in nineteenth-century science was commonly known as the primitive universal ocean. On the third day dry land emerged, plants were created, and they were continuously destroyed and renewed, creating new genera and species throughout the succeeding days.[44] Dawson placed the first Azoic rocks as belonging to the third day of creation. The arrangement of the solar system was completed on the fourth day, when the dominion of "existing causes" began. It was with the beginning of these causes that science started. To this period belong the metamorphism of Azoic rocks and the Cambrian period; the Palaeozoic and Mesozoic rocks belong to the fifth day.[45]

The whole series of the fossiliferous rocks belonged to the fifth and sixth days, and Dawson wrote that there was no geological evidence for the early plant creation of the third day, or for the geological changes on the fourth day. Early on the fifth day, many tribes of animals were created and continued to exist through the sixth day.[46] On the sixth day, the first to appear were the herbivorous mammals; then second, a variety of terrestrial reptilian and other lower forms not included in the work of the previous day; and third, the carnivorous mammals. The herbivores predominated in the earlier creations of the period, and this period corresponds with the Tertiary era in geology.[47] At the beginning of the Tertiary period, most of the gigantic reptiles of the Mesozoic period had disappeared, and terrestrial mammals of large size and high organization had taken their place.[48] As not all animals that can be found both contemporarily and in the fossil record were mentioned in the Bible, linking the periods of geology with the days of Genesis and describing the appearance of various forms of animal life were problematic. But for Dawson this problem was only apparent, as he argued, it was not necessary to assume that Moses noticed all the animals and plants that had ever lived but, instead, that he accounted of the *first appearance* of each great natural type in the animal and vegetable kingdoms.[49]

Dawson explained the existence of fossil evidence for now-extinct animal life by making two interlinked arguments: first, "man" is different from all other animals in that the original exemption from death in Eden only applied to "man."[50] Second, it was God's will that certain creations should go extinct as the environment changed. However, upon the fall from innocence, God placed a curse on the world, which allowed dangerous creatures to enter Eden. "Man" was now forced to live among creations that were originally intended to have become extinct, so that Eden could have expanded to cover the entirety of the Earth.[51] This, according to Dawson, explained both the issue of the lack of dangerous animals in Eden, as stated in the Bible, and why many animals that were contemporaneous to humans extended far back into the tertiary period—that is, they were introduced earlier on the sixth day of Genesis, and not at the same time as "man." "Geology and revelation, therefore, coincide in referring the creation of man to the close of the period in which mammals were introduced and became predominant, and in establishing a marked separation between that period and the preceding one in which the lower animals held undisputed sway. This coincidence, while it strengthens the probability that the creative days were long periods, opposes an almost insurmountable obstacle to every other hypothesis of reconciliation with geological science."[52] By correlating the geological periods with the days of creation outlined in Genesis 1–2, Dawson had extended the time line for the creation of the Earth in a way that was in accordance with geological evidence for an old Earth, while maintaining the integrity of the creation narrative. This allowed for Dawson to pinpoint the emergence of humans to a more recent time, on the sixth day of creation.

THE CASE FOR MONOGENESIS

> *The Bible, as we have seen, knows but one Adam, and that Adam not a myth or an ethnic name, but a veritable man: but some naturalists and ethnologists think that they have found decisive evidence that man is not of one but of several origins. The religious tendency of this doctrine no Christian can fail to perceive. In whatever way put, or under whatever disguise, it renders the Bible history worthless, reduces us to that isolation of race from race cultivated in ancient times by the various local idolatries, and destroys the brotherhood of man and the universality of that Christian atonement which proclaims that "as in Adam all die, so in Christ shall all be made alive."*[53]

As important as the early history of the Earth and appearance of animals was to Dawson, the real sticking point for him in reconciling the cosmogony of the Bible with the teachings of science was human history. Dawson was highly attuned to the scientific milieu in Britain and his works were full of references to concurrent scientific controversies and the newest research.[54] *Archaia* included references to figures such as James Cowles Prichard (1786–1848); Robert Chambers (1802–1871); Darwin, Miller, and Lyell; the German explorer Alexander von Humboldt (1769—1859); the theologian Thomas Chalmers (1780–1847); the physician, philologist, and ethnologist Robert Gordon Latham (1812–1888); the Swiss-born biologist, polygenist, and Harvard professor Louis Agassiz (1807–1873); and many others. In the first half of the nineteenth century a plethora of books were published that questioned the biblical-human time line and human origins or sought to reconcile the biblical explanations with evidence of an ancient Earth from science. These include books such as Prichard's *Research into the Physical History of Man*; the physician and medical lecturer William Lawrence's *Natural History of Man* (1819); Latham's *Natural History of the Varieties of Man* (1850); and the *Vestiges of the Natural History of Creation* (1844), written anonymously by Chambers. As Secord has noted, the *Vestiges* made evolutionary theories a conversation topic and domesticated the radical notion of development.[55]

In the first half of the nineteenth century as well, debate over the unity of human origins raged. In summary, the division was between those who believed humans originated from the same root (monogenists) and those who believed in the plural origin of races (polygenists). Dawson was a monogenist and an anti-transformist.[56] He believed that various types of life had been created during the different days in Genesis and that all humans originated from Adam and Eve.

Dawson considered the Bible to be very clear about the unity of Man. Thus, if it were true that "man" was not of one origin, but that he came from multiple origins, it would cast serious doubt on the validity of the Bible, as "the Bible knows but one species of man. It is not said that men were created after their species, as we read of the groups of animals."[57] Dawson asserted that although some naturalists advocated for polygenesis, evidence from science and philology—in addition to Scripture—overwhelmingly supported monogenesis. Having assured the concerned reader that there was no discord between evidence from Scripture and science, Dawson outlined how proper science showed that humankind was one species with multiple varieties and did not consist of multiple species.

A key issue in the debate over the origins of humans was the categories of species and varieties. Drawing on research from Georges Cuvier (1769–1832), Alphonse Pyramus de Candolle (1806–1893) and Dana, Dawson outlined four points for a definition of species. First, there is a certain range of uniform characters in a species, although the range varies with the species. Second, the interval between species has to be distinctly marked. Third, the uniform characters of species have to be constant through generations over time. Fourth, the liability to variation within a species has to be part of the description of a species.[58] All naturalists would agree, Dawson argued, that these four points are based on nature, and moreover, that they are "absolutely necessary to the existence of natural history as a science."[59] Consequently, a deviation from this view of species meant a deviation from proper scientific practice.

Dawson drew heavily on Prichard's research, and his view of species and varieties was functionally similar. Like Prichard, Dawson believed species remained constant. Efram Sera-Shriar has shown how Prichard constructed his system for understanding human difference and the unity of races by, first, using physical and cultural characteristics to organize humans of the world into a taxonomized system and, second, using this system to demonstrate that human races were varieties, not species.[60] John Cornell has further shown that, later in life, Dawson slowly relaxed his opposition to what he perceived as the fundamental flaw in theories of evolution, namely, the analogy between transmutations of varieties and species. However, in the 1850s and 1860s Dawson was in staunch opposition to such ideas.[61] Dawson's four-point definition of species was simultaneously broad and specific. It was broad so that it allowed for great variation within any given species, and specific enough that when applied against the physical characters of humans the unity of humans followed from its definition. In his discussion of varieties, Dawson outlined a six-point definition. It was an open-ended definition with the first point stating that, "the limits of variation are very different in different species."[62] Dawson argued that differences within varieties of adults of the same sex can be vast, and in many cases the liability to vary is a part of the characters of the genus. The physical environment and intelligence allowing adaptations of animals, and certain physical attributes such as the "adaptability of the digestive and locomotive organs to varied kinds of food and habitat," all affect the expression of variations.[63] When applied specifically to humans, the six-point definition of varieties were as follows:

1. Man's living situation is the most artificial of all animals. Man's structure is adaptable to all sorts of land regions.
2. Races of man do not have essential characters of species.
3. Men vary in the same ways as domesticated races.
4. Men do not vary more than what can be observed in other animals.
5. All varieties of men can intermix, and in most cases the intermixture positively affects the "vital energy and vigor" of the offspring.[64]

6. Varieties of men could have originated because of external condition, in the same way as domesticated animals.

A key figure in the debates over the unity of humans and the definition of varieties and species was Louis Agassiz.[65] Agassiz believed that humans had been created by God both as multiple species and in multiple centers of creation.[66] While both Dawson and Agassiz attributed the origins of humans to divine creation and argued for the immutability of species, here ends any similarity. When Dawson outlined his definition of species and varieties, Agassiz was the key opponent in his narrative. According to Dawson, Agassiz's position was so unreasonable that he was the only one who held this view, arguing that "the only modern naturalist of eminence who seems disposed to attempt this proof of the diversity of origin of species, is Prof. Agassiz, whose principal argument is the geographical distribution of animals."[67] This exaggeration served to make the point that when Agassiz argued for centers of separate creation he was advocating something that "no one can be fairly called on to believe."[68]

As a polygenist Agassiz argued against the idea of a single original pair of humans—such as Adam and Eve. In Agassiz's definition of varieties and species, varietal difference was made to approach species differences. This is in stark opposition to Dawson's definitions, which asserted a clear differentiation. Agassiz by contrast blurred the distinction between the two categories.[69] Dawson argued that his breakdown of the categories of species and varieties as they related to the way physical differences manifested themselves in humans, compared to other living creatures, proved the unity of the human species. There was simply not enough difference among humans to suggest that the various races were separate species and not simply varieties: "The attempt, then, sanctioned by so great a name as that of Agassiz, to establish diversity of origin for the individuals of the same species on the ground of geographical distribution, falls to the ground; and perhaps fails most signally of all in the case of man. We may, therefore, safely rest on that philosophical necessity for the unity of origin of each species with which we commenced this part of the inquiry; at least until it shall be shown that the individuals of some one true species must be diverse in origin."[70] Dawson thus asserted that the burden of evidence was on the polygenist, and not the other way around.[71]

In Dawson's estimation, it would take only one thousand years for the development of the strongest varieties of humans to occur. After outlining what is a very set definition of species and varieties, Dawson turned to the unity of the human species's singular origin, using evidence from philology. Drawing on research from figures such as Latham, Prichard, the linguist and philologist Friedrich Max Müller (1823–1900), and the diplomat Baron von Bunsen (1791–1869), Dawson argued that although humans had many and varied languages (and the diversity at first may have seemed to indicate radical racial diversity), philology showed the links between them through vocabulary and/or grammatical structure—even in languages that appeared completely different.[72] But using evidence from philology raised an additional problem: the Earth history time lines of the cultures whose languages Dawson was drawing on to show the unity of man had much longer time lines for human history than what the Bible could allow for, which suggested the existence of pre-Adamites. The key problem was determining what type of evidentiary source was valid, and which type should take priority.

SECULAR AND SACRED HISTORY

> *This requires a differentiation between scripture and secular history when they both refer to the same fact or event, and while the latter can be used as a secondary aid to understanding the former, one cannot find the whole truth in them.*[73]

What type of written account is a valid record of past events? This was a key question for Dawson to address in *Archaia* when discussing the origins of humans. He considered the Bible to be a historical record independent of its religious claims. For him the main difference between secular and sacred history was that the Genesis account referred back to a prehuman period, and thus to events humans could not have observed, as "writers may accurately relate contemporary events, or those which belong to the human period, without inspiration; but the moment that they profess accurately to foretell the history of the future, or to inform us of events which preceded the human period, we must either believe them to be inspired, or reject them as impostors or fanatics."[74] The Mosaic cosmogony, he argued, was neither a scientific account nor a "poetical myth" but should instead be considered a "direct revelation from the Creator."[75] Although sacred and secular history informed each other, sacred history fundamentally had the higher authority. Dawson constructed this argument by asserting that the Hebrew cosmogony was the only written history that did not have a mythical period.

In *Adam's Ancestors*, David Livingstone has shown the long history of attempts to reconcile the Bible, especially Genesis, with three key challenges: the first was the existence of pagan chronologies, the second was the genealogy of monstrous creatures shown in fossils, and the third was voyages and encounters with the New World. In *Archaia* Dawson countered these challenges in two ways.[76] He discredited the historical accounts of all other cultures except the Egyptian and Hebrew and made a distinction between antediluvian and postdiluvian peoples. The existence of pagan chronologies that differed from the Christian did not necessarily cast valid doubt on the biblical. Dawson considered it a particular human trait both to be interested in the past, present, and future and to be religious. This was a special attribute only present in humans and one that marked the separation between humans and lower animals, as "man as a 'religious animal' desires to live not merely in the present, but in the future also and the past."[77] Dawson argued that, although all religious systems included ideas about the future and the distant past, not all religious systems or historical accounts were made equal:

> The only ancient nations that have given us in detail their own written history are the Hebrews, the Hindoo, and the Chinese. The last people, though professedly very ancient, trace their history from a period of barbarism; a view confirmed by their physical characters and the nature of their civilization; and on this account, of no other, their history cannot be considered as of any ethnological value. The early Hindoo history is palpably fabulous or distorted, and has been variously modified and changed in comparatively modern times. The Hebrew history, as it bears on this point, we shall notice in the sequel. There is one great and very ancient people—the Egyptian—evidently civilized from the beginning of all history, that have not succeeded in transmitting to us, except in garbled fragments, their history; but have left abundant monumental evidence

> of great events that transpired among them; and, except the Hebrews, these are the only people who can profess to give us any authentic ancient history carrying us back to the origin of man.[78]

Having earlier in *Archaia* established that the authority of the Bible rested on the assertion that Genesis was inspired revelation, Dawson did not devote much space to establishing the reasons that the historical and religious narratives of other cultures were untrustworthy. Simply put, the Bible was more credible than other narratives. Dawson argued that it "has no mythical period. It treats of no ages of gods and demi-gods, claims no fabulous antiquity for its people, asserts no divine origin for its heroes."[79] According to Dawson, the Bible was a trustworthy document because, as a monotheistic religion, it accounted for one creative God, and it was especially explicit about the origins of the Earth and nature. By contrast, "the heathen" gave divine power to subordinate gods and natural phenomena.[80] Therefore, Hindu and Chinese cosmogonies were untrustworthy because they contained mythical periods and had intermixtures of the human and divine. Although the Bible also contains accounts of miracles, this was different, as Dawson argued that they were miracles performed by a monotheistic God.

In her book *James Cowles Prichard's Anthropology*, Hannah Augstein emphasized the significance of Egyptian history—for ethnology in general and for Prichard, specifically. For these early ethnologists, Egyptian history was able to reach as far back as that of the Hebrews and, thus, to the dawn of humanity. Because of its importance for ethnological research, Dawson could not simply discard the Egyptian time line in the way that he had discarded other time lines.[81] Dawson therefore had to acknowledge that Egyptian historical accounts were trustworthy. However, to acknowledge this posed a problem, because many interpretations of the Egyptian time line stretched too far back to correlate with Dawson's analysis of the Genesis account. For instance, Bunsen had estimated the date of the first king of Egypt, Menes, was around 4000 BCE, which would set the Deluge at about 10000 BCE.[82] This, Dawson argued, was impossible. Dawson modified the chronology of Menes, dating his rule back to 2000–2500 years BCE, which he aligned with the spread of humankind from Noah's three sons after the Deluge. The existence of cultural artifacts and practices that appeared older than 2500 BCE was due to remains from the antediluvian peoples that were (almost) all eradicated with the Deluge: "The Bible gives us a definite epoch, that of the deluge, for all human origins; but though no family but that of Noah survived this terrible catastrophe, it would be a great error to suppose that nothing antediluvian appears in the subsequent history of man. Before the deluge there were arts and an old civilization, and after the deluge men carried with them these heirlooms of the old world to commence with them new nations."[83]

Dawson asserted that while the Deluge was a geographically local event, for humans it was all encompassing, as it fundamentally changed human culture and the physiology and anatomy of races.[84] This was Dawson's scheme for human origins and diffusion; the cradle of life—or the location of Eden—was somewhere in Western Asia.[85] The human family split into two races early on, one that focused on the mechanical and fine arts and one that focused on the "moral endowments of the species."[86] When the two races mixed, they produced "a race excelling both in energy and physical endowments—the 'giants'—mighty men of violence."[87] If we were to find remains from these antediluvian people we would probably see them as a different and extinct species of man. After the Deluge, only Noah's family survived, and they settled in the fertile plains of the Euphrates and Tigris. Noah had three sons, Japheth, Ham, and Shem, who each formed a main stem that further diverged into several family branches as humans were scattered all over the Earth. The Japhite nations went to Europe and Northwestern Asia. The Shemite nations remained by the Euphrates and Tigris valleys and the neighboring regions. Ham's family spread over Southern Asia and Northern Africa. In the beginning, all of Noah's family had the skills to own domesticated animals and the principle arts of life. These characteristics of civilization were then lost with the more scattered tribes, who became hunters. At the time of the exodus, there was already a substantial difference in the languages of human groups. The Japhite nations especially spoke a language that was unintelligible to the other nations.[88] In Dawson's view, this showed that the Bible explained the difference in appearance and cultures of humans as they appeared throughout the world.

The notion that antediluvian humans would look very different from humans today should be seen within the context of discussions over fossil finds, debates over "missing links," and the developmentalism of books such as the *Vestiges*.[89] Dawson cautioned the reader that finds of human remains or work of art in the tertiary deposits—which Dawson equated with the early sixth day of Genesis and therefore with the early arrival of humans—might be due to "accidental intermixtures" of the deposits, as "ordinary observers are so little aware of the sources of error against which it is necessary to guard."[90] As a coping mechanism for explaining away a discord between geological discoveries and Genesis, this does not appear to be a very convincing one, and perhaps this is why Dawson only briefly touched upon the subject of human fossil remains in *Archaia*, noting simply: "I have referred to them merely to point out connecting-links between the secular and sacred history of the earlier part of the human period, as a useful sequel to our comparison of the sacred history with the conclusions of science, and as furnishing hints which may guide the geologist in connecting the human with the tertiary period, and in distinguishing between the antediluvian and post-diluvian portions of the former."[91] Distinguishing between the antediluvian and postdiluvian served simultaneously to discredit all human history narratives other than the Egyptian and Hebrew ones and to argue against those scholars who outlined a long time line for Egyptian culture. In this way Dawson proposed an interpretation of geology, biblical exegesis, and philology that was in accordance with a short human time line and a long time line for the Earth, while sidestepping (or warning the reader against) archaeological finds from voyages of exploration and discoveries made by the emerging science of paleontology.

Archaia was commercially unsuccessful, but its updated edition, *Origin of the World According to Revelation and Science* (1877), sold almost fifteen hundred copies during the first year.[92] With *Origin of the World*, Dawson moved toward a tone that was more explicitly defensive of the role of Christian theology for science. Susan Sheets-Pyenson has suggested that this was part of the reason *Origin of the World* was more popular. The title visibly juxtaposed itself against *Origin of Species* and perhaps conveyed the content of the book more clearly than the title *Archaia* had done. *Origin of the World* explicitly addressed what Dawson saw as the major problem with Darwin's work, and of those advocating his theory, which was the need for retaining the role of religion in science. This was similar to what he had done in his review of *Origin of Species* in the *Canadian Naturalist and Geologist* in 1860.[93] Dawson had not addressed the content of *Origin of Species* in *Archaia*, but the reviewers of *Archaia* made the link between the two books. For example, in May 1860 the *North British Review* noted that they would "cordially recommend Dr. Dawson's . . . as a safe guide" to *Origin of Species*.[94] By contrast, the *Saturday Review of Politics* argued that, compared with *Origin of Species*, Dawson's doctrine of the immutability of species was outdated (as would quickly become clear).[95] Dawson's staunch opposition to the theory of evolution generally became unattainable within the elite scientific milieu in the second half of the nineteenth century. As Ronald Numbers has pointed out, by the late nineteenth century he was one of the only prominent creationists in North America.[96] And yet Dawson was able to continue to disseminate his scientific research program from a position of power, and his influence within the elite scientific scene did not wane. He was able to maintain both his views on human history and evolution, on one hand, and his influence on science and society on the other. Dawson is an example of how our views on the relationship between science and religion, the content and nature of scientific practice, and the extent to which science was secularized in the nineteenth century can be further complicated by looking outside of Britain and to the imperial world.

Archaia was the beginning of a long series of publications directed toward a nonspecialist reading audience on the subject of science and religion. It represented a first full exposition of Dawson's view on the relationship between science and religion, and how this related to the history of the Earth and human history. Dawson sought both to direct the research programs of scientific practitioners and religious scholars in Canada and to shape perceptions of geology and human history as being in accordance with the Bible for a general reading audience. The choice to go with a Canadian publisher, which he developed a strong relationship with, very likely gave him more control over what he could publish compared to the British publishing world, where his ideas, from an elite scientific point of view, were already beginning to be outdated by the time he published *Archaia*.

Dawson historicized humans by drawing on evidence from both science and the Bible. He developed a day-age theory that plotted the geological periods onto the days of the Genesis narrative. In this way, he sought to maintain a time span for the Earth's geological history that was in accordance with current scientific ideas, while also establishing a short time frame for the emergence of human life in accordance with a more literal reading of the Bible. This allowed Dawson to keep what he considered to be the integral parts of the Genesis narrative and show that geology was by no means a threat to Christianity. A key part of constructing this system was his rejection of polygenism and the origin narratives of other cultures. He criticized the advocates of polygenism, in particular Agassiz, for their use of what he argued was a circular definition of species and varieties, because it placed the plurality of humans as part of the premise. Yet he similarly outlined a definition of varieties that was so open-ended that a species could vary, seemingly indefinitely, without breaking off into a new species. His definition of species and varieties assumed that species remained permanent and nonevolving. With regard to humans, Dawson drew a large part of his evidence for monogenism from philology. His treatment of the evidence from philology, which drew on languages from cultures that had radically different origin narratives from the Judeo-Christian one, was situated within a conceptual framework that prioritized the Hebrew and, to some extent, Egyptian accounts of the early history of humankind over those of other cultures such as Hindu and Chinese narratives. These were discredited as untrustworthy and clouded in myth.

Historicizing humans in this way was an attempt by Dawson to circumvent the growing body of literature that extended the human time line beyond the biblical—as well as the literature by the polygenists—and guide the nature of Canadian science and society. Dawson believed that careful studies of science and the Bible were mutually beneficial. Studying the book of God and the book of Nature was important not only for the advancement of knowledge but also for the moral status of humans. For Dawson, both geology and scripture showed the progressive character of creation. He believed that God had not created everything like it was at present and that the world would continually change in the future. As the only creation, Dawson argued, humans were cast in God's image, but it was up to humankind to live up to this original intent. Humans have the option of either sharing in the future glory of the renovated condition of the planet or sinking into endless degradation. He believed that the careful study of science and the Bible would help humanity in achieving an elevated moral state in preparation for the future of humankind—or, the eighth day of creation. It was a key part of the vision he had for Canadian science and society. When Dawson historicized humans in *Archaia*, he did so both to discredit the theories of polygenism and developmentalism and to save the souls of his readers.

3

HISTORICIZING BELIEF

E. B. Tylor, Primitive Culture, *and the Evolution of Religion*

EFRAM SERA-SHRIAR

During the 1860s the scientific naturalist and ethnologist-turned-anthropologist Edward Burnett Tylor (1832–1917) began producing his foundational work on the history of human culture. His first book, *Anahuac: or, Mexico and the Mexicans Ancient and Modern* (1861), was a travel account of his voyage through Mexico during the mid-1850s. It contained a mixture of his own personal observations on the historical and contemporary state of the indigenous population of Mexico and testimonies of other travelers who had visited the region since the Spanish Conquest.[1] Tylor's next major book, *Researches into the Early History of Mankind* (1865), traced the historical development of human culture.[2] It laid the groundwork for what would become his most significant contribution to British anthropology, his two-volume magnum opus, *Primitive Culture* (1871).[3]

At its core Tylor's *Primitive Culture* was an evolutionary study that aimed to trace the developmental history of cultural attributes among the races of the world. One of the major theories that he outlined in the work was "animism"—the idea that all religions evolved from a rudimentary belief in spirits animating the world. By locating the laws that governed the development of religion, Tylor attempted to plot all forms of worship onto an evolutionary scale showing how religious beliefs transformed from basic understandings of the world as being animated by spirits to complex religious systems such as Christianity. Central to his purpose was an attempt to naturalize all religions and explain their ontologies using scientific theories. He was not trying to reconcile science and religion but, rather, to bring religion under the domain of scientific understanding.

This chapter will explore how Tylor historicized religious beliefs in *Primitive Culture*. In particular, it will look at the way he constructed his theory of animism by exploring the various influences that shaped his work both within Britain and throughout the empire. Although many scholars have discussed aspects of Tylor's research in the secondary literature, he still remains underexplored. He was a significant contributor to the development of British anthropology as a discipline, yet there has been only one major biography, written by his former colleague and friend Robert Ranulph Marett (1866–1943) and published in 1936.[4] A more extensive examination of his work, therefore, sheds important new light on the impact of his theories and methods for the discipline of anthropology. This chapter builds on the scholarship of historians such as George Stocking, Robert Kenny, and David Livingstone, who have explored human history theories in nineteenth-century scientific works, as well as Frank Turner and Bernard Lightman, who have looked at the contest of cultural authority between science and religion in Victorian Britain. Furthermore, it connects to a large body of secondary literature by scholars such as Janet Browne, Michael Bravo, Daniel Carey, Peter Hulme, and Tim Youngs who have examined the importance of travel literature in the making of scientific knowledge.[5] This study diverges from previous examinations of *Primitive Culture* by bringing together these three disparate areas of historiographical analysis and showing how human history theories, the relationship between science and religion, and travel narratives all shaped Tylor's writings.

Although he was forwarding an explanation of religious beliefs using a type of evolutionary theory, Tylor's work was not embroiled in any major conflict with either the Church or any religious figure. This was a marked difference from how the writings of other scientific naturalists such Thomas Huxley (1825–1895) were received by religious communities during the same period.[6] Tylor's engagement with the intersection of science and religion was fascinating in other ways. Despite his clear aim to replace religious explanations of the world with scientific ones, a close examination of his book exposes a complex and strenuous relationship between science and religion. Many of the sources he used to exemplify extra-European religious practices came from the travel reports of missionaries and other types of travelers with strong religious convictions. Tylor was reliant on these firsthand accounts to substantiate the credibility of his writings. Even though he wanted to push religion to the margins of ethnological and anthropological research, he was unable to avoid religiously influenced sources for his data. This raises some interesting questions about the boundaries between the so-called two spheres of science and religion.[7]

This chapter is divided into two main sections. The first part provides an overview of Tylor's career prior to the publication of *Primitive Culture*, paying particular attention to the importance of his journey through Mexico during the 1850s in shaping his later ethnological and anthropological writings. In the second part the focus shifts to his writings in *Primitive Culture*, and it examines the types of sources that influenced his evolutionary theory of religion.

EDWARD BURNETT TYLOR: EXPLORER, ANTHROPOLOGIST, AND CULTURAL THEORIST

Anthropologists have canonized Tylor (see fig. 3.1) as the father of modern anthropology, referring to the discipline in the nineteenth century as "Tylor's science."[8] Some scholars have claimed that his contributions form the basis of many of the cultural theories still used by researchers today. Tylor has been depicted as the great teacher of civilization. He has been credited as the first practitioner to define the term "culture," a pioneer in sending his students into the field for ethnographic data, and a forerunner in devoting his career strictly to the scientific study of human culture.[9] Tylor's influence on anthropology was already being recognized during the second half of the nineteenth century and into the early twentieth century. At his seventy-fifth birthday in 1907, his former student and friend Andrew Lang (1844–1912) argued that "he who would vary from Mr. Tylor's ideas must do so in fear and trembling (as the present writer knows from experience)."[10] Tylor's writings are representative of what has come to be known as the "cultural turn" in Victorian anthropology. From the 1860s onward the discipline experienced a major shift in focus that saw a decrease in studies that prioritized physical descriptions of races to an increase in studies that emphasized cultural descriptions.[11] Some of the major cultural studies to be published during this period were *Ancient Law* (1861) by the legal scholar and historian Henry James Sumner Maine (1822–1888), *Primitive Marriage* (1865) by the lawyer John Ferguson McLennan (1827–1881), and *Pre-Historic Times* (1865) by the politician, archaeologist, and entomologist John Lubbock (1834–1913).[12] Tylor was also one of the key figures to champion this epistemological movement, and a short survey of his early life affords an opportunity to examine some of the key stages that contributed to the development of his work as an anthropologist and proponent of human cultural history.[13]

FIGURE 3.1. Photograph of Edward Burnett Tylor with his signature, circa 1870s. Reproduced by permission of the Pitt Rivers Museum, University of Oxford (catalogue no. 1998.266.1).

Tylor was known for his racial sensitivity, which was something embedded into his psyche from a young age. He was born into a middling-sort Quaker family and received most of his education from Grove House School in Tottenham, which was operated by the Society of Friends. Nineteenth-century Quakerism was a culture immersed in abolitionist causes. Most members of the Society of Friends were committed to religious and racial tolerance, and they were inclined to be relativistic and sympathetic toward people who held different beliefs from their own, as well as people who had other racial backgrounds. Many leading ethnological figures in the first half of the nineteenth century were Quakers, including the physicians James Cowles Prichard (1878–1848) and Thomas Hodgkin (1798–1866). Their religious beliefs influenced several of the discipline's central tenets.

The Quaker doctrine of the "inner light," for example, placed great emphasis on viewing all humans as equal, and this aligned easily with monogenesis—a theory postulating a common human ancestral origin.[14] By highlighting physical and cultural similarities between all races, monogenetic theories supplied Quaker ethnologists with scientifically grounded arguments against the exploitation of extra-Europeans and generated support for racial equality through moral, physical, intellectual, and social improvement.[15] Tylor's immersion in a social milieu where issues of racial parity were at the forefront of communal discussions shaped his later work on human culture. Throughout his long career, he was a staunch monogenist, and in his *Researches into the Early History of Mankind* and his *Primitive Culture* he forwarded arguments in favor of human unity.

Tylor's initial interest in ethnology and natural history, however, came from his older brother, Alfred Tylor (1824–1884), who was an archaeologist, geologist, and brass founder. According to Stocking, it was Alfred who introduced Tylor to ethnological topics and persuaded him to go abroad. After the death of his parents in 1852, Tylor began working in his family's foundry and within a few years he developed tuberculosis, which forced him to change careers. Alfred encouraged his younger brother to visit North America to clear his lungs, and taking his brother's advice Tylor set out on a two-year trip in 1856. These travel experiences represented a formative period in his ethnological and anthropological writings, because it was during this trip that he engaged for the first time with extra-European cultures.[16]

In the introduction to his travel narrative, *Anahuac*, Tylor wrote that he spent "the best part of a year" traveling down the Mississippi River and observing the Native Americans and African slaves he encountered along the way. He also lived for a short time on a sugar plantation in Louisiana before deciding to visit Cuba for a new adventure.[17] While in Havana he met the ethnologist, archaeologist, and banker Henry Christy (1810–1865). This chance encounter was an important juncture in Tylor's life, as it was through his friendship with Christy that his interest in race studies blossomed. There was much in common between Christy and Tylor. Both were Quakers, grew up in London, and came from middling-sort families. Moreover, they had similar educational backgrounds, having studied at schools operated by the Society of Friends. With so much in common, they quickly established a strong rapport. Christy would become Tylor's mentor over the course of the next ten years until his death in 1865. He would teach Tylor how to observe ethnological subjects in situ, as well as explain the major tenets of Prichardian monogenism. When the two first met in 1856, Christy invited Tylor to accompany him on a four-month horseback journey through Mexico.[18]

Before the 1820s the Spanish crown heavily guarded information about the interior of their colonial landholdings in the Americas. There was widespread fear in Spain that their European rivals would attempt to steal their territorial possessions if data about their resources was made available. Non-Spanish travelers from Europe required permission from Spanish authorities before they could embark on any form of exploration within the region. The Spanish also imposed rigid trading rules in their colonies, and between 1739 and 1748 the British were at war with Spain.[19] As a consequence of this sociopolitical climate, few reports of Mexico by British writers existed prior to the collapse of Spain's imperial rule in the Americas. However, things began to change at the onset of the nineteenth century, and between 1804 and 1825 the region was at war. The various colonial states in the Americas were vying for independence from the Spanish crown. In the aftermath of these conflicts, it became easier for British exploration teams to investigate the interior of places such as Mexico.[20]

Tylor's description of his journey through Mexico was a valuable resource for British ethnology because there were so few accounts of the country written by English-speaking authors. Most ethnological studies of Mexicans relied on older Spanish reports for ethnographic data, and many British researchers believed that these accounts were filled with inaccuracies and exaggerations.[21] One of Christy and Tylor's main objectives during their excursion was to rectify this problem, and they collected as much ethnographic and archaeological information as possible on the contemporary and historical condition of Mexicans. Upon returning to Britain, this material could then be organized, analyzed, and reworked into ethnological studies. Tylor's narrative, therefore, represented an important contribution to British ethnography, because it was one of the most up-to-date British accounts of Mexico available. Tylor emphasized in his introduction to *Anahuac* how he and Christy had immersed themselves in Mexican culture, basing their analytical understanding of the society on "local knowledge and experience."[22] By examining the details of his Mexican journey we can begin to see how Tylor developed his later views on the history of human culture.[23]

From the opening pages of his travel narrative Tylor noted that Mexico was a remarkable location for conducting ethnological investigations. He argued that the substantial amount of archaeological evidence available throughout the countryside made it possible to trace the history of the indigenous peoples from the earliest stages of their civilization to the present. He did not just look at materials from his saddle as he passed through an archaeological site; he would dismount from his horse and engage directly with the objects. As an example, when he visited the pyramids near Micaotli he wrote that he "sat cross-legged on the ground" while members of the local community "brought many curious articles in clay and obsidian" for him to examine.[24] He recognized that his primary experience traveling through Mexico and seeing the landscape through his own eyes reshaped his understanding of the region's past. Most notably it afforded him an opportunity to correct the reports of other travelers who had visited the region, because he could verify, alter, or reject their accounts based on his own personal experiences.

We can also see the beginnings of Tylor's fascination with extra-European religions in *Anahuac*. It was an interest that grew over the next decade, and that contributed to the formulation of his theory of animism in *Primitive Culture*. Throughout his travels in Mexico, he recorded observations on the religious practices of the indigenous population. In some of his ethnographic descriptions we see the seeds of his later comparative studies of religions. As an example, he observed that some aspects of Christianity bore resemblance to Aztec rituals. He wrote, "Practically, there is not much difference between the old heathenism and the new Christianity. . . . They had gods, to whom they built temples, and in whose honor they gave offerings, maintained priests, danced and walked in processions."[25] Because there were similar forms of religious rites present in modern Christianity, this suggested to Tylor that there was a potential common ideological foundation to all belief systems. This was a theoretical line of inquiry that Tylor would return to in his later anthropological work.

In other instances, he recounted occasions of seeing indigenous people appropriating and reinterpreting European beliefs into their own culture. Tylor, as an example, relayed an incident in Mexico City where he and Christy came across a festival during the holy week of Easter. The indigenous people were buying pasteboards in the shape of Judas, which they set afire and blew up with gunpowder as part of their celebrations. This practice, according to Tylor, readapted an older Aztec custom of setting alight evil idols, into a Catholic ritual of burning Christ's betrayer. This was not an unusual activity, and many Mexican traditions in Tylor's view mixed aspects of the two cultures.[26] There were other ways in which European and indigenous customs blended together in Catholic settings. Tylor related an occasion where he and Christy visited a monastery and convent in the mountains near Cacahuamilpan and watched a group of youths dressed in traditional Mexican clothing perform European-style dances for the nuns and monks. He wrote, "The costumes were evidently intended to represent the Indian dresses of the days of Montezuma . . . and, to our unspeakable astonishment, [they] began to dance the polka. Then came a waltz, then a schottische, then another waltz, and finally a quadrille."[27] In his later anthropological work, Tylor would explain this blending of older cultural practices into newer ideologies as "survivals," which were antiquated understandings of the natural world that were influenced by the remnants of primitive thinking in modern societies.[28] During his journey through Mexico in the 1850s, however, he was still developing his cultural theories. Thus, there was no explicit mention of cultural survivals in his travel narrative.

Tylor was also engrossed in studying the religious relics of Aztec civilization, and he included many observations on their material culture in his writ-

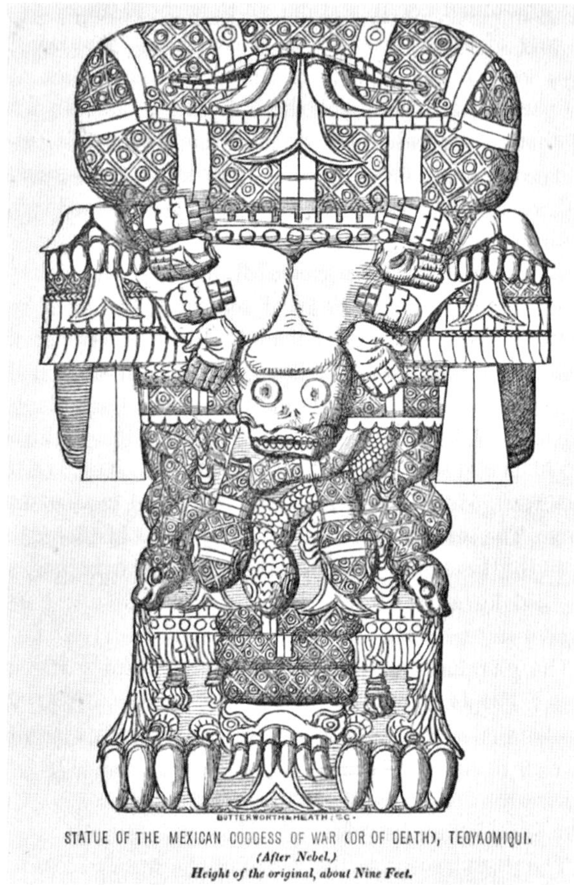

FIGURE 3.2. Woodcut of a stone totem of an Aztec war goddess. From Edward Burnett Tylor, *Anahuac: or, Mexico and the Mexicans, Ancient and Modern.* (London: Longman, Green, Longman, and Roberts, 1861), 221. Author's collection.

ing. This interest would later inform some aspects of his theory of animism.[29] When visiting a museum in Mexico City, Tylor described in detail a large Aztec totem: "The stone known as the statue of the war-goddess is a huge block of basalt covered with sculptures. The antiquaries think that the figures on it stand for different personages, and that it is three gods,—Huitzilopochtli the god of war, Teoyaomiqui his wife, and Mictlanteuctli the god of hell."[30] Tylor even included a woodcut of the totem (see fig. 3.2) in his narrative for his readers to utilize in their investigations. It was a fairly detailed representation of the original artifact that illustrated the ornamental nature of the totem, with its skulls, snakes, and monstrous feet. Tylor's engagement with artifacts such as the totem of the Aztec war goddess was one of the many examples of him documenting and interpreting the religious history of Mexicans in *Anahuac*.

Tylor's travels through Mexico had a significant effect on his later ethnological and anthropological writings for several reasons. First, by traveling to Mexico and seeing the indigenous peoples firsthand, Tylor could claim an authoritative understanding of ethnological subjects and collect substantive data upon which to base his research claims. Second, his time in Mexico under the guidance of his mentor, Christy, was a kind of intensive practical training course in ethnography. Third, Tylor came to value firsthand observations because it meant that the researcher was not reliant solely on secondary accounts, which often misrepresented or misunderstood aspects of extra-European cultures because of a lack of personal familiarity with different races. This was one of the reasons that Tylor utilized a wide range of travel accounts in *Primitive Culture* to substantiate its credibility as a scientific text. Fourth, it was through his extended interactions with indigenous populations in Mexico that Tylor began taking an interest in the religious beliefs and practices of extra-European peoples. Finally, his ethnographic writings on Mexicans were entrenched in historical analysis, a methodological approach that would endure in his later studies of extra-European cultures.

When Tylor returned to Britain in the late 1850s, his status quickly rose within the ethnological community. By the early 1860s he was becoming a leading figure at the Ethnological Society of London (founded in 1843), holding close ties with both Huxley and the younger generation of scientific naturalists, who were becoming increasingly more active in scientific studies of race, and with Christy and Hodgkin, who were members of the older group of Prichardian monogenists.[31] Tylor even had connections to the physician and speech therapist James Hunt (1833–1869), and the Anthropological Society of London (ASL), serving as the society's foreign secretary from 1863 to 1864 before (according to Stocking) he was enraged by Hunt's "pugnacious racism," which "offended his humanitarian Quaker beliefs."[32] Tylor did not join Huxley and Lubbock in their attack on the ASL in the 1860s, however, and because he distanced himself from these debates he was not a target of Hunt's staunch criticisms of the ethnological community. This could possibly explain why, during the aftermath of the anthropological schism in the early 1870s, Tylor was able to continue to build upon his first-rate reputation; his name was not directly associated with either camp.[33]

Because his family owned a successful brass foundry, Tylor was able to devote himself entirely to the study of ethnological and anthropological topics. It was not necessary for him to find gainful employment because his family's wealth could support his research. This enabled him to produce many important works throughout the second half of the nineteenth century, including *Researches into the Early History of Mankind*, *Primitive Culture*, and his textbook *Anthropology* (1881).[34] Even with his financial independence, he ultimately settled for a life in academia. By 1884 he was appointed Keeper of the Natural History Museum and Reader in Anthropology at Oxford. Tylor remained at this historic institution for the remainder of his career. Notwithstanding all of these remarkable accomplishments and contributions to the research field, it was his book *Primitive Culture* that cemented his status as a world-renowned anthropologist. The book represented the maturation of his cultural theories.

SCIENTIFIC NATURALISM, *PRIMITIVE CULTURE*, AND THE EVOLUTION OF RELIGION

As he worked through his ideas Tylor produced several earlier versions of his theory of animism, which was an integral component of his cultural evolutionary model. The first and probably most notable early incarnation was in his *Researches into the Early History of Mankind* from 1865. The following year, he published a short paper titled "Religion of Savages" in the *Fortnightly Review*. He also presented a lecture at University College London in May 1869 titled "Spiritualistic Philosophy of the Lower Races of Mankind," and finally he presented a paper at the Ethnological Society of London on extra-European religious beliefs, which was published in 1870.[35] As he refined his views on the evolution of religion, his own religious perspective changed. Tylor was increasingly drawn to rationalism and positivism as the 1860s progressed, and by the time he published *Primitive Culture* he had gravitated away from Quakerism. He officially left the Society of Friends in the summer of 1864. Although he never openly called himself an agnostic, the theories that he articulated in

Primitive Culture, as well as his association with other prominent agnostics such as Huxley, suggest that he adopted aspects of this ideological movement into his research program and worldview.[36]

Tylor became a scientific naturalist during the 1860s, and in *Primitive Culture* he wanted to explain human cultural evolution through natural causes.[37] He wrote, "The world at large is scarcely prepared to accept the general study of human life as a branch of natural science. . . . To many educated minds there seems something presumptuous and repulsive in the view that the history of mankind is part and parcel of the history of nature, that our thoughts, wills, and actions accord with laws as definite as those which govern the motion of waves, the combination of acids and bases and the growth of plants and animals."[38] All aspects of human culture, according to Tylor, were governed by natural laws, just like any other phenomena in nature. His insistence on establishing anthropological research along naturalistic lines was part of his program to secularize the research field and push religious explanations of humanity's past to the margins of the discipline. By reconfiguring anthropological discourse in this manner, human history and culture would become solely scientific topics.

Tylor's method for naturalizing anthropology, however, differed from those of other scientific naturalists such as Huxley. During the 1860s Huxley saw himself as a scientific reformer, and he was keen to introduce scientific naturalism into ethnology and anthropology. His strategy for reforming most natural sciences followed a similar process. First, he would introduce Darwinian evolutionary theory into the discipline. This was fairly straightforward in the case of ethnology and anthropology, because many practitioners within these two fields already argued in favor of common descent, and topics relating to species and varieties were at the forefront of ethnological and anthropological discussions.[39] Second, Huxley would attempt to make the science ideologically neutral by trying to separate its research from both religion and politics. It was an effective strategy for achieving an influential social status in Victorian society, because Huxley could emphasize the important and impartial perspectives the sciences had on a broad range of issues.[40] Third, Huxley would vilify researchers in the discipline who did not adopt an evolutionary stance and scientific neutrality. He targeted established scholars who clung to older ideologies, which he viewed as antiquated. For example, Huxley framed the biologist, anatomist, and paleontologist Richard Owen (1804–1892) as representing an obsolete version of naturalism, indicative of early nineteenth-century practices.[41]

Tylor's strategy for transforming the research program of anthropology included both the second and third components of Huxley's disciplinary reform tactics, but Tylor did not promote a Darwinian mandate. This was the primary difference between their two forms of scientific naturalism. In fact, there was scarcely a mention of Darwin in *Primitive Culture*—an absence that many readers noted. Tylor explained his reason for excluding Darwinian theory from his work in the preface to the second edition of *Primitive Culture* from 1889. He remarked, "It may have struck some readers as an omission, that in a work on civilization insisting so strenuously on a theory of development or evolution, mention should scarcely have been made of Mr. Darwin . . . whose influence on the whole course of modern thought on such subjects should not be left without formal recognition." Tylor continued by stating, "This absence of particular reference is accounted for by the present work, arranged on its own lines, coming scarcely into contact of detail with the previous works of [this] eminent philosopher."[42] Tylor's evolutionary theory of human culture is best characterized as a secularized version of Prichardian monogenism mixed with Comtean developmentalism.[43]

As was the case with Prichardian monogenists, Tylor was committed to an argument in favor of a shared human ancestral origin—though he rationalized his monogenetic theory through the use of natural laws—and not because of a commitment to biblical Scripture, as was the case with Prichard's writings.[44] For instance, Tylor stated that "in studying both the recurrence of special habits or ideas in several districts, and their prevalence within each district, there come before us ever-reiterated proofs of regular causation producing the phenomena of human life, and of laws of maintenance and diffusion according to which these phenomena settle into permanent standard conditions of society, at definite stages of culture." Similitude among the races was evidence in support of human unity, and he wrote, "To general likeness in human nature on the one hand, and to general likeness in the circumstances of life on the other, this similarity and consistency may no doubt be traced, and they may be studied with especial fitness in comparing races near the same grade of civilization."[45] According to Tylor this resemblance in all facets of human culture, was visible at every stage of civilization. If one were to examine the daily activities of different tribes throughout the world, one would perceive a tremendous amount of uniformity in different groups.[46]

In order to trace accurately the cultural history of humans, Tylor argued, it was optimal to produce an all-encompassing overview of every race. Only then would researchers have a full view of humankind and of its development through successive cultural stages. His definition of culture was broad; he described it as "that complex whole which includes knowledge, belief, art, morals, law, custom, and any other capabilities and habits acquired by man as a member of society."[47] His anthropological paradigm was truly global, and it utilized data from all over the world. Like his ethnological forebears Tylor depended on the observations of travelers to substantiate the credibility of his research. Monogenists such as Prichard, Hodgkin, and the surgeon, Arctic explorer, and cofounder of the Ethnological Society of London Richard King (1810–1876) saw travel reports as among the most reliable forms of ethnographic data to utilize in their studies, because people who had engaged firsthand with extra-European populations had recorded them. Therefore, these accounts were given a primal status within the discipline. Tylor was building on this methodological approach in *Primitive Culture.* He had also come to appreciate the value of direct observation from his experiences traveling through Mexico during the 1850s. Nevertheless, like any form of data, travel narratives could still contain inaccuracies and misinformation. To ensure their quality, ethnologists and anthropologists needed to test them.[48]

The problem was that many travelers lacked formal training in scientific observation, and they did not possess a background in either natural history or medicine—two fields that were fundamental to ethnological and anthropological research during the nineteenth century.[49] In order for a travel narrative to be credible for ethnological or anthropological purposes, each account had to be considered on a case-by-case basis. The trustworthiness of an author had to be assessed. Was this a superficial observer? Did the author possess local knowledge of the peoples' languages and customs? Were the observers prejudiced in any way? These were important questions that Tylor argued "every ethnographer ought to keep clearly and constantly before his mind." One way to circumnavigate these concerns was by comparing the reports of multiple travelers who had visited the same region over several generations. Tylor wrote that the ethnographer "is bound to use his best judgment as to the trustworthiness of all authors he quotes, and if possible to obtain several accounts to certify each point in each locality."[50] If certain characteristics were recorded repeatedly in the narratives of a specific race, those qualities could be deemed as representative of the population. If one could trace the trait over several generations through the use of both contemporary and historical reports, it was possible to historicize these features. If similar attributes were perceived among neighboring groups, it was also possible to link these peoples to a common origin. Travel narratives, therefore, were key evidentiary resources for building monogenetic frameworks.[51]

Once Tylor had decided which sources contained reliable information on the races of the world, he used the data to construct his evolutionary theory of human culture. Prichardian monogenism may have provided him with a framework for determining human unity, through a comparison of similarities among the races, but Comtean positivism gave him a mechanism for explaining cultural development over time—especially with regard to the way different races understood the world around them.[52] As was the case with Auguste Comte's social developmental theory, Tylor argued that humans progressed through three stages: savagery (distinguished by hunting and gathering), barbarism (distinguished by pastoralism and agriculture), and civilization (distinguished by industrialism).[53] As races became more civilized they also become more secular, replacing their religious beliefs with rational scientific thought.

For Tylor all cultural paradigms were essentially the legacy of previous generations, readapted into new circumstances. The task for the ethnographer was to understand how human culture transformed through each successive stage, by looking at how different technologies, habits, customs, and belief systems changed over time and appropriated new meanings. If there was insufficient data available on a specific race because of a lack of historical record, analogies had to be drawn with other groups. Under this model, "savage" races were seen as analogous to prehistoric Europeans. The two groups could explain one another and even suggest how "savage" cultures might progress based on European history. The comparative method was a core feature of Tylor's developmental theory.[54]

Tracking cultural evolution was a complex task, and Tylor conceived of various concepts in order to follow different transitions in societies. One of his principal notions was the idea of "survivals," which he described as "processes, customs, opinions, and so forth, which have been carried on by force of habit into a new state of society different from that in which they had their original home, and they thus remain as proofs and examples of an older condition of culture out of which a newer has been evolved."[55] A prime example of cultural survivals were religions, which he viewed as archaic understandings of nature, whose evolution could be traced through successive stages of civilization as human groups altered and improved their knowledge of the world. Because Tylor was in essence arguing that all religious beliefs stemmed from savage thought, he foresaw that his readers might view his writings as controversial.

To prevent a potential outcry from his audience Tylor maintained that, when studying the evolution of religion, ethnographers had to keep an objective mind and separate themselves from religious dogmatism. He stated, "The connexion which runs through religion, from its rudest forms up to the status of an enlightened Christianity, may be conveniently treated of with little recourse to dogmatic theology. The rites of sacrifice and purification may be studied in their stages of development without entering into questions of their authority and value, nor does an examination of the successive phases of the world's belief in a future life demand a discussion of the arguments that may be adduced upon it for our own conviction."[56] Despite trying to marginalize the authority of Christian theologians, Tylor claimed that his attempt to historicize religions was uncontroversial. It was not intended as an attack on or a critique of religious thinking per se. Rather, as Tylor asserted, his theory was to be strictly seen as a scientific examination of belief systems that was based on objective observations and not personal biases. He argued, "In these investigations, however, made rather from an ethnographic than a theological point of view, there has seemed little need of entering into direct controversial argument, which indeed I have taken pains to avoid as far as possible."[57] By asserting that he was being above board about the implications of his anthropological investigation, Tylor hoped he could evade any sort of clash with either the Church or a religious figure. It was a highly qualified approach to naturalizing the human sciences.[58]

The next step for Tylor in accepting his evolutionary theory of religions was acknowledging that there was much to be gained by studying "savage" beliefs. He wrote, "Far from its beliefs and practices being a rubbish-heap of miscellaneous folly, they are consistent and logical in so high a degree as to begin, as soon as even roughly classified, to display the principles of their formation and development."[59] According to Tylor all humans were naturally predisposed to animistic thinking, and the most rudimentary forms of religion were organized along cogent principles. The ethnographer could follow their development as they transitioned into ideologies more akin to modern European belief systems. Moreover, understanding these transformations would enable ethnographers to determine the roots of religion and explain why humans accepted the existence of God in the first place. Tylor warned his readers that any study of religion had to contain a fairly broad definition of what constituted a belief system. If it were too specific, it would eliminate many forms of belief that were essential to understanding the underlying reasons for religiosity. His base line for tracing religious development was his theory of animism—the idea that at the core of all religions there is the notion that spirits (or supernatural forces) animated the world. If researchers were to accept this concept as the basis of their investigations, it would provide them with a much more inclusive framework for studying extra-European religions.[60]

Tylor devoted the entire second volume of *Primitive Culture* to the study of animism. Each chapter analyzed different aspects of religious thinking through a global perspective, tracing its development through the three stages of civilization. Most of the focus was on savage thought, because Tylor believed that every religious practice observed in Christianity stemmed from primitive cultures. Chapter 8, as an example, examined religious rites and ceremonies. Tylor wrote, "As a contribution to the theory of religion, with especial view to its lower phases as explanatory of the higher, I have here selected for ethnographic discussion a group of sacred rites, each in its way full of instruction, different as these ways are. All have early place and rudimentary meaning in savage culture, all belong to barbaric ages, all have their representatives within the limits of modern Christendom."[61] He divided these rites into five groups: prayer, sacrifice, fasting, orientation, and lustration. For his examples he used the observations of travelers.

What is interesting about his use of travel narratives for ethnographic purposes is that he used many accounts written by religious writers such as missionaries. He could not completely remove these types of reports from his study because in many instances only missionaries had visited some of the more remote human populations around the world.[62] Despite trying to push religious figures out of scientific race studies, Tylor was still reliant on some of their observations for his data. The lines that he drew between science and religion became blurred as a result. He did not draw attention to this problem in *Primitive Culture*, but it still raises interesting questions about the relationship between science and religion during the nineteenth century, because many of the authors that he cited in his work were heavily influenced by their faith. Although he was not explicit about this in his writings, Tylor's emphasis on his own objectivity was central to his attempt to reconcile this issue. It did not matter if the source was shaped by the author's religious convictions, as Tylor could claim that his professed neutral ideological stance removed any potential biases.

There were other ways that Tylor tactfully justified the use of religious sources in *Primitive Culture*. When writing, for instance, on the prayer rituals of Tannanese islanders (in modern-day Vanuatu), he wrote: "In the Papuan Island of Tanna, where the gods are spirits of departed ancestors, and preside over the growth of fruits, a prayer after the offering of first-fruits is spoken aloud by the chief who acts as high priest to the silent assembly: 'Compassionate father! Here is some food for you; eat it; be kind to us on account of it!'"[63] His citation shows that the description came from the narrative of the Scottish missionary Rev. George Alexander Turner (1818–1891), who had traveled to Polynesia in 1840 and spent nineteen years exploring the region. Turner had a comprehensive knowledge of the different island communities and even collected a large amount of Polynesian artifacts for museums in Britain.[64] He was therefore perceived as a reliable ethnographic collector. Tylor could ignore any latent issues that Turner's religious orientation potentially produced because his protracted stay in the South Pacific, combined with his familiarity of the indigenous populations, cemented his ethnographic observations as an invaluable resource.

In another example, Tylor discussed the prayer rituals of Samoans, and he stated, "When the libation of ava was poured out at the evening meal, the head of the family prayed thus:—'Here is ava for you, O gods! Look kindly towards this family: let it prosper and increase; and let us all be kept in health. Let our plantations be productive; let food grow; and may there be abundance of food for us, your creatures. Here is ava for you, our war gods! Let there be a strong and numerous people for you in this land.'" Tylor again cited Turner's narrative as a trustworthy account, but he also cross-referenced it with the writings of the English missionary William Ellis (1794–1872) who had also witnessed similar practices when he resided in Polynesia between 1816 and 1824.[65] Because both figures recorded analogous religious rites in Samoa, Tylor reasoned that they were likely representative of the islanders' prayer rituals.

The reports of British missionaries in Africa also furnished Tylor with data on extra-European religions. For instance, when examining the prayer practices of indigenous communities in South Africa Tylor wrote, "The Zulus, addressing the spirits of their ancestors, think it even enough to call upon them without saying what they want, taking it for granted that the spirits know, so that the mere utterance 'People of our house!' is a prayer." His source for this information was the travel narrative of the English missionary Henry Callaway (1817–1890), who worked for the Church of England.[66] Callaway, unlike other missionaries such as Turner and Ellis, possessed some scientific knowledge that was useful for ethnographic purposes. This added more credibility to his observations. Callaway had received his surgical license in 1842 and his apothecary license in 1844. Many surgeons found employment in colonial settlements or on voyages of exploration as medical officers. High demand for scientific information from around the world meant that they were often requisitioned for natural history collecting, which included ethnographic reporting. For these reasons, surgical students acquired some training in natural history during their schooling. Callaway was no exception, and he was thus primed for ethnographic research. Given his background, it would have been easy for Tylor to justify Callaway's inclusion in *Primitive Culture*. Callaway possessed fairly typical attributes for natural history collection.[67]

Not all of the religiously influenced sources in Tylor's writings were from the reports of missionaries. He also relied on the observations of other types of travelers with religious convictions. For instance, when describing lustrations ceremonies among New Zealanders (cleansing rituals akin to baptisms), Tylor substantiated his analysis through a combination of sources. He verified their credibility by cross-referencing them and showing that they contained similar descriptions. The first account was by the English missionary Richard Taylor (1805–1873), who sailed to New Zealand in the 1830s.[68]

The second was by the Jewish merchant Joel Samuel Polack (1807–1882), who had resided in New Zealand between 1831 and 1837. During his time in the South Pacific, Polack recorded detailed descriptions of the cultural attributes of the New Zealanders. He published his account, *Manners and Customs of the New Zealanders*, in 1840.[69] Polack is an interesting example because his Jewishness was in some ways an asset. When he arrived in New Zealand, the other European settlers ostracized him because of his religious beliefs. Only the indigenous peoples were willing to engage with him, and as a result he formed a strong rapport with them, giving him access to more facets of New Zealander life than were normally afforded to Westerners. Moreover, having lived most of his life among Christians, Polack was also accustomed to observing the habits and mannerisms of social groups that were different from his own. He was attuned to studying the peculiarities of different cultures, which made him a key informant for anthropological researchers such as Tylor.[70]

Tylor's method for historicizing religious beliefs in *Primitive Culture* was part of his larger framework for understanding the development of human societies. Religion, like all aspects of civilization, was part of a cultural legacy inherited from previous generations. As human groups progressed from savagery to civilization, these cultural attributes were readapted into new circumstances and worldviews. In constructing his developmental model, Tylor used the standard ethnological approach of comparing historical and contemporary travel accounts to demonstrate uniformity among the races. This in turn was then employed as evidence in favor of monogenesis. His historicizing of human cultural traits did not stop there, and after establishing a common origin for all races, based on similarities perceived throughout the cultures of the world, his next step was to trace how the most rudimentary forms of religious thought (e.g., the belief in spirits animating the world) evolved into complex religions such as Christianity. To do this, he adopted the positivistic model of Comte and combined it with the comparative method, so that he could draw analogies between the belief systems of primitive tribes, barbaric nations, and civilized societies. Tylor's theory of animism was inherently critical of all forms of religiosity. Yet, despite the secularizing argument in *Primitive Culture*, many of the sources that Tylor used in his analysis were written by figures with strong religious convictions.

For the most part, reviews of *Primitive Culture* were positive.[71] In May 1871, an anonymous reviewer for the *Athenaeum* described it as "singularly clear in style, rich in learning, [and] methodological in arrangement." The reviewer continued by stating that Tylor's book bore "ample witness to the fact that much honest and intelligent labor has been bestowed upon it."[72] *Primitive Culture* received an equally glowing review in the *Examiner*, which commended Tylor for his masterful treatment of the evolution of religion: "To trace philosophically the history of spiritual beings from the soul of man upward to the Great Deity is the no slight task that the author has set himself. In its performance he has worked out, in a new field, and with the best results, that fertile conception of development or evolution." Admiration for Tylor's work did not stop there, and the reviewer wrote: "He has done for the spirit world what Mr Darwin has done for the animal."[73] In other words, Tylor had created a naturalistic theory for explaining the origins of religiosity and the laws that governed its development.

Even in instances where the review was written by a figure with strong religious convictions, there was little criticism of Tylor's secularizing theory. For example, the Scottish Presbyterian minister and philosopher Henry Calderwood (1830–1897), reporting for the *Contemporary Review*, wrote, "This new work, bearing the title of 'Primitive Culture,' is by far the most important contribution which the British press has yet given to the departments of anthropology and ethnology."[74] Calderwood also praised Tylor for his substantial use of traveler reports for corroborating his anthropological theories. Like Tylor, Calderwood believed that the most reliable ethnographic observations were recorded by figures that had spent significant periods of time living among extra-European populations. In his review he listed several ethnographic observers, whom he deemed trustworthy and vital sources for anthropological studies, including the British explorers David Livingstone (1813–1873), Samuel Baker (1821–1893), and John Hanning Speke (1827–1864); the orientalists William Gifford Palgrave (1826–1888) and Richard Francis Burton (1821–1890); the naturalist Alfred Russel Wallace (1823–1913); and the missionaries Ellis, Turner, and John Smith Moffat (1835–1918). Because Tylor used data from all of these figures' accounts in

Primitive Culture, Calderwood argued that the book was an invaluable and well-referenced text.[75]

By 1872 Tylor's book was being reviewed in other countries, and it continued to receive strong approval. A journalist for the *Calcutta Review* stated that Tylor had "given to the world a very valuable addition to the particular department of ethnology which he cultivates, in these volumes. He has made himself, by an enormous amount of labor and care, one of the most reliable authorities regarding the primeval customs and beliefs of mankind." The reviewer asserted that *Primitive Culture* was of particular value to Europeans living in India because they had "so many opportunities of becoming acquainted with the customs and beliefs of tribes comparatively low in the ranks of civilization and of adding to the amount of available knowledge."[76] Thus, Tylor's book had tremendous value for the empire, because it helped colonial settlers better understand the extra-European cultures around them. In addition, it provided Europeans living abroad with a general sense of what types of information were valuable for anthropological research.

In the same year an anonymous writer for the *North American Review* applauded Tylor for his work in *Primitive Culture*. The reviewer stated that there were "few erudite treatises which are at once truly great and thoroughly entertaining." The author continued by comparing Tylor's book to Charles Darwin's *Origin of Species* (1859), arguing that "we do not know how we can more aptly express our sense of the thoroughness of Mr. Tylor's scientific culture than by saying that he constantly reminds us of the illustrious author of the 'Origin of Species.'"[77] Much like Darwin, the reviewer believed, Tylor had articulated a comprehensive and naturalistic framework for understanding why cultures transformed over time.

Yet not everyone agreed with this opinion. In his review of *Primitive Culture* that appeared in the *Academy* in 1872, Alfred Russel Wallace criticized Tylor for not providing an explanation of the causes that shaped cultural development. He stated that Tylor had exhaustively classified ethnographic "facts" (e.g., examples of different cultural practices from sources such as travel accounts) without elucidating the phenomena that produced them. He was also unhappy with Tylor's description of the modern spiritualist movement in Britain as representing a survival of primitive belief.[78] It is interesting that one of the most negative reviews of Tylor's book came from a fellow scientific naturalist, and not a theologian or other type of figure with strong religious convictions. Nevertheless, Wallace was not critical of Tylor's secularizing theory of human culture. His more probable motive for criticizing *Primitive Culture* was that Tylor was ridiculing his spiritualist beliefs.

To conclude, in this chapter I explored how Tylor historicized religious beliefs in *Primitive Culture*. In particular, I looked at the way he constructed his theory of animism, by exploring the various influences that shaped his work, both within Britain and throughout the empire. By bringing together into conversation three major historiographical themes (human history theories, the relationship between science and religion, and travel literature), I aimed to shed important new light on Tylor, the disciplinary history of British anthropology, and the secularization of science in the nineteenth century.

What emerged was an interesting story about the complex relationship between science and religion. Although Tylor was trying to push religion to the margins of anthropological research, he was unable to do away with it completely. Missionaries were some of the few ethnographic observers to have spent significant time living among some of the more remote extra-European populations around the world, and their writings were essential to Tylor's anthropological research program. Thus, his attempt to fully secularize the evolutionary study of religions was unsuccessful. When Tylor historicized religious beliefs in *Primitive Culture*, he was ultimately indebted to Christian missionaries as well as other types of travelers with strong religious convictions for his data. Such an analysis further challenges extreme versions of the conflict thesis, showing the subtleties of the relationship between science and religion in the nineteenth century. While there has been quite a lot of work written on the conflict thesis by historians of science, very little has specifically discussed the conflict thesis in relation to anthropology. It is the point of this chapter, therefore, to encourage future investigations into this important historiographical dialogue. After all, in the wake of Tylor's writings on the anthropology of religion, there was a new generation of researchers interested in historicizing human beliefs, including two of the more notable examples, Edward Clodd (1840–1930) and James George Frazer (1854–1941). Both would become hugely influential in the research field and would build upon the foundational writings of Tylor's studies in the anthropology of religion.

4

THE HISTORY OF THE "RED MAN"

William Bollaert and the Indigenous People of the Americas

MAURIZIO ESPOSITO AND ABIGAIL NIEVES DELGADO

Before the travels of Alexander von Humboldt (1769–1859), the Spanish and Portuguese colonies were mostly unknown to northern European explorers. The Iberian countries were very protective of their colonies and wanted to control any strategic information that could jeopardize their monopoly. Humboldt himself requested a formal permit from the king of Spain, Charles IV (1748–1819), in order to visit the colonies, aware of the Spanish government's general distrust for foreigners with scientific ambitions in the New World. As is well known, Humboldt's romantic descriptions of the Amerindian flora, fauna, and civilizations inspired young naturalists and travelers—including Charles Darwin—to undertake arduous journeys to the Americas. But for many younger explorers following in Humboldt's footsteps after the 1810s, the aims and interests were different from the romantic ideals that had characterized the previous generation.[1]

Indeed, while Humboldt was publishing and advertising his best seller in Europe, important political events were happening in the Americas that would dramatically change the Iberian dominions' relations to the Old World. From the 1810s the colonies became gradually independent. The process left important regions open for foreigners seeking new profitable enterprises.[2] Now freed, Argentina, Peru, Chile, Ecuador, and other new political entities became gold mines for entrepreneurs. An overflow of northern European travelers of various shades, including explorers, naturalists, and entrepreneurs, arrived in South and Central America. Most of them were more interested in commercial opportunities than in the dramatic landscapes or the "savages" that older voyagers had previously described. Mary Louise Pratt has dubbed these postcolonial explorers the "capitalist vanguard," an assorted group of explorers mainly attracted by mining prospects and economical ventures.[3]

One of these explorers was William Bollaert (1807–1876), a young chemist who had worked with Humphry Davy (1778–1829) and Michael Faraday (1791–1867) at the Royal Institution. Born in 1807 in Lymington, Hampshire, Bollaert could not count on any aristocratic background for his livelihood. The blindness of his father interrupted his medical career, and in 1825 he sailed to Peru in order to work as a chemist and mineral assayer near Tarapacá (in modern Chile).[4] Bollaert would spend several years in the Americas. Traveling through many regions from Patagonia to Texas, he became a distinguished and well-respected anthropologist, geographer, and linguist of the Americas. As a correspondent for the University of Chile, the secretary of the Anthropological Society of London, and a member of the Ethnological Society of London and the Royal Geographical Society, Bollaert had many opportunities to disseminate his ideas and investigations in Britain and elsewhere. Despite the vast literature he produced on many different subjects (more than eighty articles and three monographs), he remains relatively unknown within the history of British exploration, imperialism, and anthropology. The aim of this chapter is to fill this gap, if only partially.

However, the chapter does not cover all of Bollaert's interests and undertakings. More specifically, the aim is to describe, assess, and contextualize his anthropological ideas on the indigenous inhabitants of the New World. In particular, this study explores Bollaert's defense of polygenism as it relates to his extensive studies on cultural traits including language, customs, archaeology, geography, religion, and culture.[5] Indeed, to Bollaert, historicizing humans did not mean reconstructing the phylogeny of the human race; it meant recognizing the historical circumstances differentiating the communities of indigenous peoples living throughout the Americas. The fact that indigenes did not speak the same language or share the same religions, customs, or traditions implied that there were different histories to be told for each group. Bollaert contributed to the task of documenting these differences through his work as a correspondent and ethnographic collector, who provided information, ethnological objects, curiosities, and natural history specimens for scholars and museums in the United Kingdom and other countries.

Although Bollaert was never a key figure in the nineteenth-century debate on the origin and nature of "man," he is historically interesting for three principal reasons. First, in following his journeys we are introduced to an underworld of scholars, entrepreneurs, politicians, and dealers of different kinds that have not received the attention they should, at least in relation to the production of ethnological and natural history knowledge within the context of nineteenth-century Latin America. As Michael Bravo has argued, it was fundamental for the ethnologists in Britain to have access to reliable information coming directly from people living or traveling in the British colonies and elsewhere.[6] The Victorian ethnological and anthropological compendia were possible thanks to the large network of people collecting specimens and information in situ.[7] Figures like Bollaert were crucial mediators between the metropolitan naturalists working comfortably in their libraries and the dangerous worlds overseas. If many Victorian armchair anthropologists valued highly the firsthand information coming from the field, it is important to pay more attention to these travelers in order to have a deeper understanding of how nineteenth-century anthropological knowledge was produced.[8] We think that, in moving away from the big names of the history of science who are too often at the forefront of the secondary literature, we are in a better position to appreciate the vast web of people and interests involved in the production of certain kinds of knowledge. This also allows us to move beyond the well-known historiographical categories of "armchair theorist" and "fieldworker" in anthropology.[9] In fact, Bollaert was both an armchair theorist and a fieldworker. He was collector, explorer, and antiquarian proposing his own theories about who the indigenous peoples of the Americas were and whence they came.

Second, by examining the work of Bollaert we are better positioned to understand the context in which particular ideas and beliefs emerged; how questions about human origins and nature acquired different meanings and justifications in diverse places and situations, for different subjects and according to different interests. Indeed, the singular experience of Bollaert in Latin America reveals how the curiosity for the human past was intimately connected with a host of larger political preoccupations shared by Creoles and Europeans. For instance, the idea of the category of the "Red Man" having its separate origin from other races owes part of its rhetorical force to these postcolonial sites dominated by imperial interests and civilizing dreams. The peculiar style of constructing the "native" that we find in Bollaert's work, as well as in those of other British "vanguard capitalists," responded to the necessity of assessing those who could be part of the civilizing agenda and those who could not.[10] The indigenous peoples' good or bad nature, their humanity or inhumanity, their civilized or uncivilized manners, their articulate or inarticulate language, their laziness or industriousness, their origins—in other words the nature of the "Indian"—were all established against the values and interests of the hasty observer, who had to establish whether a territory could be colonized, exploited, cultivated, industrialized, and so on for the progress of European imperialism and knowledge. Indeed, as David Livingstone has put it, "The question of human origins was not simply about science and species; it was about society and sex, cultural identity and racial purity. Bloodlines mattered—economically, politically, and spiritually."[11] We think that Bollaert's writings render visible the network of interests and ideas making the category of the "Red Man" convincing and useful on both sides of the Atlantic.

Third, Bollaert is interesting because his writings clearly show how ethnologists from the first half of the nineteenth century historicized humans before the physical racialization of the anatomists and physicians.[12] In following the sixteenth-century Jesuit missionary José de Acosta (1539–1600) and other earlier "ethnologists," Bollaert considered that cultural differences such as language, religion, astronomy, architecture, politics, and social organization were fundamental to understanding the origins of and relations among races. Thus, in line with sixteenth-century Spaniard missionaries and naturalists, the Bollaertian "Red Man" not only defined a class of human beings with supposedly homogenous morphological traits,[13] but it also defined a large population sharing some identifiable cultural patterns. Bollaert believed that the history of the "Red Man" was not merely visible in the faces, skin color, or the form and size of the skulls; it was also visible in the architecture, languages, religions, myths, and art of the people. In accounting for human differences, Bollaert distinguished between race and species. For instance, the "Red Man" was a single species, containing several races. Although Bollaert maintained that the "Red Man" was unrelated to other human species (it was a special creation), the racial variations that the explorers could recognize in the different regions of the Americas had to be ascribed to the local environment and contingent histories. In short, different unrelated human species exhibited intraspecific racial variations that were historically produced. Thus, Bollaert historicized humans from a cultural viewpoint but dehistoricized their deep biological past.

IMPROVING THE NEW WORLD

From the late eighteenth century, as Richard Drayton argues, British imperialism was characterized by a rhetoric of "improvement." He wrote, "Central to the new species of British imperialism which emerge at the end of the century is the idea that colonization was an enterprise of amelioration. . . . Empire was now a process of preparing the rest of the world to become fully human."[14] In concert with the prospect of amelioration, the new imperialism also responded to the needs and goals of Western industrialization—a process requiring raw material, new products, and cheap manufacture.[15] These historiographical views help us contextualize Bollaert's writings within the language of "improvement" and "entrepreneurship." The new imperial enterprise consisted of a civilizing mission that replaced the old divine providentialism with a more secularized endeavor of progress through European technology, politics, and commerce. Bollaert agreed that the new rising independent nations of Latin America might be bettered through a European-inspired reorganization, but he was not convinced that the indigenous people could be improved. In line with the old theological discussions following the Spanish conquest, where knowledge about the nature, origins, and behavior of the Amerindians was explicitly connected with the possibility of Christianization and the legitimacy of Spanish imperial appropriation, "capitalist vanguards" like Bollaert believed that knowing the origins and nature of the "Red Men" meant assessing whether indigenous people could participate in the European civilizing process or be subjugated and marginalized for their own good (and for the profit of colonists).[16] For Bollaert and many other nineteenth-century travelers, two of the central questions were who and what could really be ameliorated and modernized.

In general, when we consider the accounts of most of these nineteenth-century travelers, we realize they shared very little with eighteenth-century accounts of the noble savages living in a pristine world. They shared even less of the magnificent visions of Humboldt, the emotional style of Mungo Park (1771–1806), or the descriptive sophistication of Darwin. Jean Franco harshly characterized the ideals and intentions of some of these explorers. Referring to the British travelers in South America at the beginning of the nineteenth century, she argued that these explorers were "missionaries of capitalisms and their aim was no other than the informal colonization of the continent."[17] The use of the word "missionary" is particularly fitting. Indeed, while the old traditional missionaries wondered whether the indigenous people could be good Christians, the modern ones wondered whether the indigenous people could be good citizens and workers. As a consequence, missionaries of capitalism such as Bollaert had to invent a new conception of the Americas as a region in need of development, progress, and massive European immigration. After all, supposedly local "savages" could not be improved from their natural state. Bollaert's fundamental justification for the irremediable inferiority of the indigenous populations of the Americas was their separate origins. These views were not only shared among these travelers but also diffused by the educated local elites, who often supported British ambitions. While British travelers saw Spanish America as a blank slate for political reform and economic progress, many Creoles regarded these explorers as new missionaries of modernity, who would bring the benefits of science, technology, and capitalism to backward and superstitious people.[18]

Two of Bollaert's references—the Argentinean writer and statesman Faustino Sarmiento (1811–1888), and the Ecuadorian naturalist, Manuel Villavicencio (1842–1925)—demonstrate the presence of this ideology of improvement among Creole thinkers in Latin America. Sarmiento, in the first chapter of his magnum opus, *Facundo o Civilizacion i Barbarie en las Pampas Argentinas* (1845), lamented that the "American races live in laziness and are incapable, even through constriction, of systematic dedication to hard work. This suggested the idea of introducing blacks into America, which have produced even worse results." Sarmiento instead praised the Scots and Germans who, with hard work, had thrived. He wrote, "It gives compassion and shame in the Republic of Argentina to compare the German and Scottish colonies in the south of Buenos Aires with the villages formed in the interiors," which, for Sarmiento, were formed of mestizos and indigenous peoples. Sarmiento was one among many Creoles who believed that there could be no future for barbarous races.[19] The latter could not be educated or ameliorated because they could not be assimilated into the civilized races. It is interesting to note that for Britons and Creoles alike, human nature was often assessed in terms of the people's capability for working and producing.

Villavicencio—who had helped Bollaert in collecting information on the characteristics and value of lands in Ecuador for British investors and immigrants—also had a negative opinion of the indigenous population of Ecuador.[20] In his 1858 monograph, *Geografia de Ecuador*, Villavicencio argued that, "among the three races that peopled the country, the white is more learned; followed by mestizo, the black and very little the Indian race."[21] Villavicencio believed that the indigenes belonged to one unique American race that was divided into several families, which were distinguishable physiognomically, culturally, and linguistically.[22] The American race in all its different families was the least literate, civilized, and the most besotted. To Villavicencio, civilization arrived after independence, as Spanish colonization had been brutish and violent, leaving most indigenous people with limited knowledge. He argued, "The savage nations such as Jivaros, Zaparos, Angueteros had the absurdist fetishism."[23] It was imperative, for the interest and future of the new Ecuadorian republic, that European investments and new colonizers arrived. Villavicencio wrote, "Our scarcity of population is the real origin of our misfortunes that cannot be cured with law enforcement."[24] In nineteenth-century Latin America, European immigration was essential for the learned and powerful Creoles.[25] It was thought that, in order to progress and thrive, the new republics had to improve their racial constitution and therefore escape from their uncivilized past.

The ideological views of Sarmiento or Villavicencio were only part of a larger trend. As David Rock observed, "In many parts of Latin America after independence, foreign imperialists became redundant as the indigenous liberals pursued imperialist agenda with uncompromising zeal and conviction."[26]

These new Creole elites in Latin America—that is, those receiving and helping foreigners such as Bollaert—had a very positive attitude toward Europeans and were, at the same time, very negative toward those they labeled as "Indians." Bollaert himself recalled a conversation he had with Creole friends in southern Peru: "My friends said, 'shall we who are *Hombres blancos*—white men—tolerate their *tirania*, they who are Indians? No, never.' I had the temerity to ask, 'Which white men?' they replied, *Nosotros, Hombres blancos*—we, the white men!" Bollaert ironically dubbed the Creole elites as: "the soi-disant 'Hombres blancos' of Spanish America," because he did not consider them as being genuinely white.[27]

Bollaert's conversation with South American Creoles shows the political dimension of the racial distinctions in Latin America. Indeed, knowledge about the origin and nature of the indigenous people was not something that just emerged from "neutral" observations in the field or from mere imperialist bias or ideology; it was something negotiated in these "contact zones," where the "Indian" was conceived of as one separate species, although with distinct qualities and temperaments. Bollaert uses various descriptors in his book *Antiquarian, Ethnological and other Researches in New Granada, Equador, Peru and Chile* (1860), including "coward," "weak," "pacified," and so on. These qualities and temperaments were often in contradiction with what was seen as the necessary qualities for civilizing humans. In fact, the image of the "Indian"—descriptions, accounts, and literature of the indigene—was directly connected to the ideology of transformation and modernization of these emerging nations, which was an ideology that both travelers and Creoles shared.

However, Bollaert and many other naturalists were aware that it was not enough to say that the indigenous peoples of the Americas were different from the "Hombres blancos"; it was also imperative to explain why this difference could not be attenuated, and why the "Red Man" could not be domesticated or assimilated into other, more civilized races. The prospect of amelioration required a deeper and more general knowledge of who the indigenes were, and whence they came. Thus, historicizing the "Red Man" and assessing the differences and continuities among the indigenous communities throughout the Americas could not be dissociated from the political environment in which Bollaert traversed.

HISTORICIZING THE "RED MAN"

I will call the native inhabitant of the New World the Red Man, to distinguish him, as far as color is concerned, from the white man of Europe, the brown of India, and the Negro of Africa.[28]

Bollaert dedicated a fairly extensive literature to what he dubbed the "red species." Although it is not clear why Bollaert came to name the indigenous people of the Americas the "red species," he probably shared a perspective that was popular at the time. Influential ethnologists such as James Cowles Prichard (1786–1848), William Lawrence (1783–1867), and Robert Gordon Latham (1812–1888) held to similar classifications of humans, which were based on the work of earlier naturalists such as Georges Cuvier (1769–1832), Carl Linnaeus (1707–1778), Georges-Louis Leclerc, comte de Buffon (1707–1788), and Johann Friedrich Blumenbach (1752–1840).[29] In the third edition of his work *On the Natural Variety of Mankind* (1795), Blumenbach in particular identifies five varieties of humans: "Caucasian," "Mongolian," "Ethiopian," "American," and "Malay," and he describes the "American" variety as "copper-colored." In this work Blumenbach also outlined the human divisions proposed by other authors such as Gottfried Wilhelm Leibniz (1646–1716), Buffon, Linnaeus, Immanuel Kant (1724–1804), Johann Georg Ritter von Zimmermann (1728–1795), Christoph Meiners (1747–1810), and Georg Simon Klügel (1739–1812). Blumenbach concluded that the color of the skin of the "American" variety was not constant, yet not so diverse "as the other descendants of Asiatic autochthones, who peopled the ancient world."[30] This conviction was not only held by Blumenbach; it was shared by a large number of naturalists and explorers. Following this view, Prichard, Lawrence, and Latham all described the American indigenous populations as being copper-colored, or red. Bollaert's work, therefore, followed an ethnological tradition that can be seen as representative of the work developed at the beginning of the nineteenth century in Britain.[31]

Ethnologists from the beginning of the nineteenth century relied on the analysis of language, institutions, material culture, and anatomy to study and classify human diversity.[32] This approach was pursued by researchers in Britain and France within the associations established in both countries such as the Ethnological Society of London, cofounded in 1843 by Richard King (1810–

1876) and Thomas Hodgkin (1798–1866), and the Société Ethnologique de Paris, founded by William Frédéric Edwards (1777–1842) in 1839.[33] Keeping with this tradition, Bollaert believed that the study of architecture, language, art, science, and religion, along with the moral, mental, and physical characteristics of the "Red Man," were fundamental to understanding the origins of the different civilizations found in the New World, as well as the relationship between them.

At the beginning of his book *Antiquarian, Ethnological, and other Researches*, Bollaert explicitly stated that he had followed the "methodical instruction on the subject of ethnology" from the *Manual of Ethnological Inquiry* published by the British Association for the Advancement of Science (BAAS) in 1852.[34] A similar manual was published in the *Journal of the Ethnological Society of London* two years later, and it was "little more than a translation of those [questions] issued by the Paris Ethnological Society."[35] This is an example of the influence of French ethnology on its British counterpart, the former being developed fifty years earlier. French ethnologists had produced ethnological manuals as early as the beginning of the century. The continuous production of ethnographic manuals and guidelines demonstrates the increasing interest in "raising the scientific criterion of its research by standardizing its methods for acquiring data."[36]

The *Manual of Ethnological Inquiry* was prepared by Thomas Hodgkin and Richard Cull, in collaboration with a subcommittee of the BAAS.[37] This manual was used as a guiding resource for missionaries, travelers, consuls, explorers, and residents to obtain accurate knowledge on people under threat of extinction. It was inspired by Prichard's address to the BAAS on the extinction of human groups, which was read in 1839. In this paper he noted the "irretrievable loss to science if so many tribes of the human family are suffered to perish, before those highly important questions of physiological, psychological, philological, and historical character . . . have been investigated."[38] Bollaert's work, then, can be seen as a contribution to this larger enterprise, connected with the epistemological framework of the Ethnological Society of London.

A second reference helping us to situate the work of Bollaert within the ethnological tradition is his explicit reference to the book *Man and his Migrations* (1852) by Robert Gordon Latham, who primarily followed a Prichardian model.[39] Like Prichard, Latham believed that physical and cultural attributes were fundamental in identifying human varieties. It was important then to refine classificatory practices and teach how to observe "in a specialized way." Following Blumenbach, Latham defined three main human varieties. However, he believed that the Malay and the American varieties were better described as subtypical modifications.[40]

Bollaert's perspective diverged in an important way from the works of the British ethnologists mentioned above, however, because he was a polygenist. The comparison between the characteristics found in the "Red Man" and the European led Bollaert to conclude that the former was "a distinct species or creation." In this way, he was closer to the work of the American physician Josiah Nott (1804–1873), whom he saw as an authority on the anatomical study of races. Bollaert wrote, "If we examine him [the 'Red Man'] by his crania and its contents, we find, at least in the conformation of the brain as compared to the European, a marked difference. On this point I cannot do better than refer to Dr. Nott's researches in *Types of Mankind*."[41] This kind of literature relied on traditional racial classifications such as those established by Linnaeus and Blumenbach and empirical investigations where variation among races is described. The analysis of physical differences was complemented with philology, iconography, paleontology, and comparative geography proving that each human race had different origins. The work of well-known polygenists such as Robert Knox (1793–1862), Samuel George Gliddon (1799–1851), and Josiah Nott, were central to Bollaert's ideas on race, physical difference, and reproduction.

It is important to mention that most of Bollaert's ideas on the origins, geography, and nature of the indigenous peoples of the Americas were published and diffused in the decades between 1840 and the late 1860s. The debate between polygenists and monogenists was especially intense in this period in both North America and Britain. During the 1850s polygenists were particularly numerous and vocal, so Bollaert's considerations were timely and widely shared.[42] But the arguments and observations Bollaert proposed for supporting polygenism were often rooted in personal experiences throughout the Americas. In several memoirs and essays that he dedicated to the topic, he repeated that the study of "man" included three principal aspects: the physical, moral, and mental.[43] All three aspects indicated that the "red species" was different from any other human types, because there was no profound continuity between the physical, moral, and mental characteristics of the "red species," and those of Negros, Asiatics (browns and yellows), or whites.[44] In 1863 Bollaert observed "the red or copper-colored men of the New World to be of one species of the genus *Homo,* a species in the scale of intelligence peculiar to itself. I have come to this conclusion by a study of their various personal characteristics, their language—without alphabets—forms of government, works of art of more than one period."[45]

The peculiarity of the "red species" required independent study, because the "Red Man" was governed by special laws.[46] Bollaert also believed that the "red species" was not homogenous. As a consequence, a thorough study of local varieties was essential because the species was constituted by a family of culturally distinct communities. Indeed, although the "Red Man" differed from other human species, there were important intraspecific dissimilarities among indigenes. Following de Acosta, who had emphasized in the sixteenth century that the word "Indian" did not refer to a homogenous class of individuals, Bollaert highlighted the important divergences across Amerindian groups: "It has been said that to see one nation of Americans you see all; this is not quite the case even as regards color; while as to form, feature, physical and mental development, there are marked differences and peculiarities resulting from causes we shall have to investigate."[47] To Bollaert there was one species, the "Red Man," but there were many races within this group.

North American races differed from Mexicans, and Peruvians were different from Fuegians in Tierra del Fuego. For instance, the Araucano race was different from the Fuegian race, and this could be proved through the study of language. Thus a characteristic such as "euphony" could provide information about the relationship between the two. As Bollaert wrote, "D'Orbigny thought the Fuegians were of the Araucano race; but the euphony found in the Araucano language is not in the Fuegian, which is very guttural."[48] To Bollaert and many early ethnologists, languages were considered to be "documents of human history."[49] The absence of "euphony" in the Fuegian race indicated a separate development between these two neighboring races. Bollaert believed that the causes making "Red Men" different were essentially historical and were related to the geographic place and way of life. He stated, "Some years since, having collected and examined materials concerning the history of America, particularly as to the architecture of ancient Mexico, Central America, New Granada, and Peru, the language spoken there, the arts, and allow me to say science, I first came to the conclusion that, in the regions above mentioned . . . each showed a civilization of its own; and that other localities in those lands supported tribes, and even nations, as mere hunters in the wilderness."[50] Bollaert personally visited many regions in the Americas, from Texas to Cape Horn, and realized that communities had adapted in many different ways to their diverse environments. He believed that, although we could separate humankind into four species, the environment caused important variations in each race over time. As David Livingstone has observed, polygenists normally rejected environmental causes for racial differentiation.[51] However, Bollaert accepted some limited external influences. Specific alimentation, diseases, and history produced important changes to the morals, customs, psychology, and behavior of the indigenous peoples. The environment could impress changes on races belonging to one human species but could not transform a human species into another. In fact, there were indigenes more or less intelligent, more or less capable of civilization, more or less ugly, more or less violent or vicious, more or less fit to work, or more or less able to acclimatize to diverse environments. Native Americans from the North, for instance, were wilder than those from Central America. They were simply hunter-gatherers, and mostly nomadic, whereas in Mexico and Central and South America complex civilizations had emerged. Studies on the writing systems among the Maya and Aztecs and observations on their architecture, astronomical ideas, and archeological objects demonstrated the differences among communities. As Bollaert showed in his many works, the geography of the "Red Man" was both complicated and dynamic, especially because new populations emerged through miscegenation from Spanish conquest and African slavery.

The important differences that anthropologists noted among populations of the "Red Man," however, could not overlook a more profound continuity. The "red species" had no ties to other species, and Bollaert made some important efforts to question the theories that argued for a primordial contact with other populations. The hypothesis of an early Scandinavian colonization was dismissed by Bollaert for lack of archeological evidence. Even though some contact might have existed with northern Europeans, they had not left any visible legacy. He wrote, "I do not entirely discredit that there may have been accidental visits from Europe to the New World, or even from China and Japan, long before Columbus found America to be the wondrous barrier to his reaching Cathay and the spice islands of the east; but the strangers from the west and east have left no trace of tongue or vestige of art in America, at least as yet none have been discovered."[52] Bollaert also considered the old suggestion of Manoel Dias Soeiro (1604–1657)—also known as Menasseh Ben Israel—who had affirmed that Native Americans were descended from the ten lost tribes of Israel. Even since Soeiro's time, a large literature had developed, arguing for a Jewish connection between the Old and the New Worlds. Many possible affinities had been proposed between customs, religious beliefs, and the languages of the two continents. However, Bollaert refuted all these discussions with Humboldt's authority, and Nott and Gliddon's *Indigenous Races of the Earth* (1857). For his part, Bollaert assured readers that his examination of the astronomy, linguistics, and archaeology in Central and South America demonstrated that the "Red Man" had no other origin than the American continent. He argued, "With the polygenistic idea we suppose that the "Red Man" is a separate creation; and all that we find in regard to him, including his astronomical and other forms of intelligence, tend to such an idea."[53] In other words, the unique cultural history of the indigenous people implied—and guaranteed—their biological disconnection from other races.

In order to support polygenesis, Bollaert showed how specific cultural products were invented separately and not inherited or diffused from other centers. For instance, Bollaert argued that indigenous astronomy in Latin America had no ties to other kinds of astronomy. He stated that "their arithmetic, division of time, name of months and days, shows that their whole system was most peculiar, and, if not absolutely original, must antedate all historical times, since it has no parallel on record."[54] Bollaert hypothesized that there were three independent astronomical traditions among the indigenous people: the Mexican, the Muisca in contemporary Colombia, and the Peruvian. He aimed to prove two things. First, that indigenous astronomy was independent and unrelated to the Old World. Second, that there was no cultural transmission from North America to the south. Indeed, if we maintained that cultural transmission traveled from the north, we would be left with the absurd conclusion that a triple importation of astronomical ideas happened. Against the proponents of monogenism, Bollaert asked why there was no evidence for similarities among these traditions and, furthermore, why, if the "Red Man" was of exogenous origins, we could not find any evidence of other cultures nowadays in the Americas. He wrote, "if it still be contended that astronomy was imported, why did not the immigrants bring in the alphabet or system of writing, the art of working iron, mills, wheelbarrows—all unknown in America; or, at least, the seeds of rice, wheat, oats, barley, &c., of their respective botanical provinces? Alas! Sustainers of the unity doctrine will be puzzled to find one fact among American aborigines to support it."[55] Bollaert believed the history of the "Red Man" showed that the indigenous people were unrelated to other human species. Thus, historicizing the natives meant to exclude the possibility of a phylogenetical reconstruction of the "Indian." In other words, Bollaert used cultural history to discredit monogenism and evolutionism.

Bollaert's arguments did not pass unchallenged. In one section of the 1865 volume of *The Ethnological Journal: A Monthly Record of Ethnological Research and Criticism*, his study on aboriginal astronomy was assessed and criticized.[56] A British writer and statesman, George Cornwall Lewis (1806–1883), rejected Bollaert's idea that Latin America's astronomical traditions proved polygenism. Lewis argued that "the weak point in mister Bollaert's reasoning arises from his not discriminating between the very different classes of facts he has collected together and their differing value for the purpose of proving his conclusions." According to Lewis, if astronomy had to be used for proving or disproving monogenism, then Bollaert was forced to the absurd conclusion that in the Old World all races shared a common origin, and in the Americas this could not be the case. In fact, Bollaert himself supposed that there was one kind of astronomy in the Old World. For Lewis, "This would be a *reductio ad absurdum* from an anthropological point of view. For we may say that there are greater affinities between some of the white races and some of the redskins, than between the whites and either the black or even the yellow races of mankind."[57] There is no evidence suggesting that Bollaert responded to Lewis. Yet it would be easy to show the fallacy of Bollaert's arguments supporting polygenism. The central point of this chapter, however, is to unravel why people like him were so obsessed with gathering evidence to question the unity of humankind. The answer cannot be found in the inner logic of the argument or in the evidence available, but in the context in which Bollaert moved.

THE INTER-CROSSING OF THE "RED MAN"

Although Bollaert was not deeply interested in the anatomical or physiological discussions related to race classification, he did consider some of the important works in the field. For instance, he relied on polygenist scholars such as Nott, Gliddon, and Knox. In 1854 Nott and Gliddon published *Types of Mankind*, a book Bollaert used as a source of inspiration in his writings. Nott had been a disciple of the famous American physician Samuel Morton (1799–1851) and dedicated the book to him. From the 1840s Morton's *Crania Americana* (1839–1849) had become a classic text for the polygenist cause. Morton's argument was simple and clear: human races were essentially characterized by differences in skull sizes. The dimension of the skull signified intellectual abilities, and the bigger the brain the smarter the person.[58] Nott and Gliddon followed Morton's idea that there was a relationship between skull size and intellectual capacity. However, in eight hundred pages, Nott and Gliddon extended Morton's findings. They believed that there were diverse zoological provinces. Each province was occupied by a differently created race. There were no ancestral relations among races, and those who believed the contrary were scientifically naïve and religiously illiterate.

Bollaert considered Nott and Gliddon's later work, *Indigenous Races of the Earth*, as "one the most valuable anthropological contributions we have as yet in our language."[59] He lamented that Knox's *Races of Man* (1850) was not mentioned enough among ethnologists, but he did not really discuss this scientific literature much in his own writings. He believed that anatomical details had to be left to "the giant of physiological science."[60] This is not surprising. Bollaert was more linguistically oriented in his ethnological research, and he believed that language was more important than skulls in identifying racial differences. The history of humankind could not be revealed through physical characteristics alone.

There was one physiological phenomenon that interested him, however: the inter-crossing of the "Red Man" with other races. The topic was extensively treated in Knox's *Races of Man*, as well as in many other works on polygenism.[61] Bollaert was deeply concerned with the consequences of racial mixing. He shared the old belief that racial mixing brought an inexorable decline in humans.[62] He also maintained that the results of racial crossings clearly demonstrated the natural differences among human species. He explained, "from what I have seen and I have been able to collect in America,—the more particularly as regards the mixture of the European with the Indian giving rise to mestizo, the European with the Negress forming the Mulatto, and the mixture of the Negro and Indian forming the Zambo, and their breeding *in and in*,—the result does not appear to me to be of a prolific nature, or satisfactory either physically, mentally or morally." To Bollaert, if the history of the "Red Man" had anything to do with biology, it was with the creation of inferior types of human beings. And yet, he added that the physical, moral, and mental defects of mixing races "would lead one to lay aside the monogenist for the polygenist view, and suppose that the white, brown, red and black, and maybe some other families of mankind, are original species; and that Mestizos, Mulattos and Zamboes, are varieties, capable only for a limited time to be prolific, while the pure species would be as persistent as ever."[63]

Bollaert historicized humans culturally, but dehistoricized them biologically. As a consequence, the infringement of natural, ahistorical, and biological laws could only be associated with racial decay. The nature of the "Red Man" could not be bettered culturally but could only be worsened physiologically through "unnatural" crossings. In short, the question about reproduction of the indigenous populations had two implications. On the one hand, the defects following racial crossings proved that the "Red Man" could not be biologically linked to other races. On the other hand, it showed that the indiscriminate crossings inexorably led to a bleak future. These two implications were especially highlighted when Bollaert discussed the works of the American archaeologist Ephraim George Squier (1821–1888). In 1854 Squier had published a monograph, *Notes on Central America*, that was based on his observations undertaken during his journeys to Honduras, Nicaragua, Costa Rica, and Salvador. Squier observed that there were two general laws in anthropology that influenced national development. The first was that, when there is intermixing between two human stocks, there is always a predominance of one over another type. As Squier argued, "nature perpetuates no human hybrids, as, for instance, a permanent race of Mulattoes." The second law prescribed that violation of racial purity produces degeneration. Squier stated, "The offspring of such combinations or amalgamations (or confusions) are not only generally deficient in physical constitution, in intellect, and in moral restraint, but to a degree which often contrasts unfavorably with any of the original stocks. The infringement of the natural distinctions among races had political effects. Indeed, anarchy, disorder, and violence in most countries of Spanish America were the result of uncontrolled miscegenation."[64] Bollaert agreed with Squier that there was an intricate relationship between anthropological knowledge and immigration policy—precisely because miscegenation produced inferior stock. The new national states had to foster new waves of European immigration so as to guarantee racial improvement. The fact that Squier, like Bollaert, was an explorer interested in economic development was no coincidence. Squier was a diplomatic representative of the United States in Central America, assessing the feasibility of an intra-oceanic railway.

The necessity to increase European immigrants in the Americas was therefore justified by the inability of indigenous groups to adapt to the modern world. Bollaert wrote, "It is very difficult to even semi-civilize the Red Man; you may modify his religious and moral ideas to some extent, but he is adverse to change from the hunter to the farmer; he loves to roam over his prairies, he hates the subjection of cities; he has occupation enough for his wants in the wilderness, and he becomes bewildered and uneasy in community with the white man. Thus he remains in the savage state, warring with his own race or revenging himself on the white intruder."[65] As with Squier and many of his other contemporaries, Bollaert was convinced that philanthropy and civilizing were useless because they clashed with the intrinsic nature of the "Red Man." Of course, there was some flexibility among the indigenous people, because the species was divided into several different races adapted to diverse environments. However, the plasticity of variation was nevertheless limited. The "Red Man" could never reach the same level of civilization as the "Asiatic Man," and the latter could not get to the apex of the "White Man." The indigenous people of the Americas were just one step up from the "Black Man." Thus, the separate origin of the indigenes explained their different (inferior) nature. Ultimately, the natural history of the "Red Man," as a naturally inferior subject, justified European plans for immigration and colonization, as well as the Aristotelian theory of natural slavery that had often justified Spanish colonization from the sixteenth century onward.[66]

Bollaert believed that human species diverged and this fact implied there was a natural order that should not be broken. The direct consequences were major and destructive; infertile or inferior herds would be produced. He argued, "Upon this sort of showing, should the different species be mixed, in the end, humanity would be destroyed."[67] He also relied on animal biology for making his point. To him, biological knowledge demonstrated that mixing species was generally harmful. In particular, Bollaert saw a revealing analogy between the supposed different races of South American camelids and human races. In a paper read at the Linnean Society in the 1830s, and updated and republished in 1863, Bollaert argued against the received view that the South American camelids, alpaca, vicuña, guanaco, and llama were related species. Bollaert reported an incident where an Australian breeder had imported a couple of specimens of each species from Chile to New South Wales.[68] The breeder's intention was to obtain better quality wool by crossbreeding. Apparently he had failed, and Bollaert used this example to argue that different species should remain distinct, because intermixing produced worse types. Even though alpacas, vicuñas, guanacos, and llamas were analogous to the four human races in Bollaert's eyes, he never mentioned whether crossing individuals belonging to the same species could improve one race.

Whereas the consequence of animal breeding was limited to the production of infertile or inferior kinds of organisms, the interbreeding among humans had important political outcomes—the impossibility of progress and civilization in Spanish America. While in North America the population of European descent had seized power without mixing with indigenous populations or Africans, in Spanish America the situation was different. Bollaert wrote, "Those of the early Conquistadores, who were not killed or died in America, returned to the mother country, often leaving children (mestizos) by Indian mothers. Then came the black African slave of both sexes. Whites, Indians, and Negros have mixed, producing endless varieties. I shall have again to refer to what has for a long time appeared to me that something detrimental, physically and morally, has being going on by mixture of the three species, since about A.D. 1500, producing at times repugnant varieties among the Zamboes, and especially from the Indian and negress."[69]

Beyond the aesthetic considerations, though, the real consequence of racial mixture to Bollaert was political instability. As Squier had argued before him, many newly independent countries in Latin America were unstable and poor because their citizens allegedly were physically deficient, mentally weak, and morally vicious because of massive miscegenation. No freedom, no democracy, no industry, and no thriving economy were possible in mixed populations. In Venezuela, Bollaert observed, a violent democracy was established; a form of anarchy without freedom because, he added, "true liberty belongs rather to the white races of Europe."[70] Other countries in Latin America were in no better shape. There was a correlation between freedom and the level of population mixture that led to despotism, disorder, and anarchy. For Bollaert, all were direct consequences of miscegenation. The only real exception, according to Bollaert, was Chile, where miscegenation was not so advanced. He wrote, "Since the separation from Spain, very many Europeans have married Chileans, and their progeny is of satisfactory character, there being no Indian or Negro blood there. The immense advantage of a country possessing a pure race can only be understood by those who have resided for a time in countries populated by mixed races."[71] Of course, these considerations were directly related to the opportunities of investment for European venture capitalists. It was not advisable to invest in mines, infrastructures, and lands in "violent democracies." At the same time, this kind of racial theorizing informed the policies of administrators in Latin American. Questions of origin, history, and race were directly and indirectly associated with the political concerns of both Creoles and foreigners.

There is no doubt that Bollaert had genuine interests in the culture of the indigenous people of the Americas. His several publications on the languages, architecture, astronomy, religion, and many other subjects attest to his real fascination with the history of the Americas. However, the interests and questions that he considered significant profoundly influenced his gaze. Humans were observed, classified, and understood as good or bad, useful or useless, functional or dysfunctional, rich or poor, and exploitable or unprofitable depending on their value to Europeans and Creoles. The nature of the "Red Man" emerged in this pragmatic space in which it was essential to establish who could, or who could not, be part of the new modern Americas, based on commerce and technology. To Bollaert, this implied that modernity could not be multiracial. The question of the origins and the history of the "Red Man" was directly related to the future of the new nations, and their possibilities for betterment. Bollaert repeatedly connected the origin and nature of the indigenes to the possibility of progress in the new emerging Latin American nations.

HISTORY, RACES, AND PROGRESS

We cannot separate the ideas of the origins, representations, histories, and descriptions of Latin American indigenous peoples from their context of emergence. While explorers arrived with prejudices and personal convictions from imperial centers, they also were immediately submerged in an environment in which similar ideas of progress and development were equally diffused. Europeans and Creoles shared the idea that the indigenous peoples were not prepared or were generally too inept to participate in the civilizing project. This was commonly justified with a polygenist narrative. The "Red Men" could not be improved because they belonged to a separate, inferior species. Historicizing or dehistoricizing humans was never a neutral undertaking, because in such contexts the question of human origins—whether they diverged from a common ancestor or were created separately—was intimately connected to the ambition to shape the Americas in Europe's image or, more precisely, in the European image of the Creoles.

Bollaert's understanding of human history and difference sheds light on one interesting aspect of the debate over human origins. Historicizing humans meant more than representing phylogenies and common ancestors. It also meant establishing continuities and differences among cultural elements. For an idiosyncratic polygenist such as Bollaert who believed that the separate races could be partially modified through the forces of the environment and human traditions, history served as a tool for disproving biological connections among races. The ruins of the old civilizations in the Americas showed that no continuity could be found with the old continent or Asia. The historical and cultural dimension of the "Red Man" made problematic the hypothesis of monogenism. But in relying on Charles Lyell and Darwin, Bollaert accepted the unbiblical antiquity of the human race as a possibility. He wrote, "we know nothing, even approximately, about how long humanity has existed on our planet. Amongst unbiblical writers of the present day, a margin is given from 35,000 to 9,000,000 years."[72] Bollaert believed that the unbiblical antiquity of the humans did not imply monogenism. Humans had a long history, but it was not a common one.

In 1876 Bollaert died in poverty in the suburbs of London. A short obituary was written in the *Journal of the Anthropological Institute* in the following year.

He was rapidly forgotten in his own country. Bollaert is mentioned only in sparse footnotes and in the specialist works of a few historians.[73] From 1860 onward the polygenists' creed started to crumble. The arguments of many naturalists and physicians for the separate creation of races, once considered very convincing and scientifically founded, gradually became considered unsound and ideologically suspicious. Monogenism triumphed together with the acceptance of Darwinism after the 1870s, although, as Robert Kenny has shown, it was no less racializing than polygenism.[74] Even though it is difficult to sympathize with historical figures such as Bollaert, analysis of his works and network shows that historicizing humans has never been a neutral undertaking. Rather, it has always been a very political one.

5

HISTORICIZING HUMANS IN COLONIAL INDIA

THOMAS SIMPSON

Man alone passes and leaves nothing. . . . Yet in India . . . we have before our eyes a society in many respects still primitive, which preserves, like a palimpsest manuscript, survivals of immemorial antiquity.

HERBERT RISLEY (1908)

This contradictory mixture of nihilistic meditation on the irrecoverable nature of mankind's past and declaration of faith in India as a pristine repository of ancient humanity appeared in a work of physical anthropology, *The People of India* (1908), by Herbert Risley (1851–1911).[1] Risley's privileging of South Asia as a location in which to discover truths about humans' deep past was premised on two claims: that the region contained "wilder tribes" that had not been wiped out, unlike many similarly "primitive" populations within the European colonial yoke; and that India's caste system was in fact a form of racial segregation.[2] His theory was contested but influential, a version fit for the heyday of physical anthropology of the long-held conviction of many European theorists that India was central to human origins and development.[3] Throughout the nineteenth century inhabitants of India—past and present, postulated and observed—were integral to understandings of humanity in Europe. Historicizing humans was also vital, both to colonial governance in India and within an era of tumultuous change in Indian political and moral thought.

The entanglement of India and theories of humanity's deep past involved competing claims as to where and who mattered in the production and transmission of knowledge. Although the nineteenth century witnessed the rise of powerful scientific institutional "centers" in Europe and North America, there was an ongoing, never-resolved series of variously influential "decenterings" of the spaces and agents of knowledge of humanity's history.[4] Wranglings over human diversity related to spatial and institutional dynamics in colonial India that were to a significant extent distinct from those in metropolitan Europe. India did not, however, operate in an epistemic vacuum. Schemes for historicizing humans crafted in or focused on the Indian subcontinent formed core elements of the most influential theories of human origins in Britain and beyond.

In the first section of the chapter, "India as the Space of Human Antiquity," I explore the notion that debates in India on human origins were particular but had wide-ranging impact on the influential scholarship of William Jones (1746–1794). Jones's work is primarily remembered today as foundational to the discipline of philology, but it was intended primarily as a comprehensive theory of human origins, which proved enormously significant during the nineteenth century throughout and beyond the English-speaking world. It was also the dominant factor in making the institution he founded in 1784, the Calcutta-based Asiatick Society (renamed the Asiatic Society in 1825), a learned institution of great prestige among European thinkers on either side of the turn of the nineteenth century.

In the second and third sections of the chapter, I develop this volume's theme of the complicated patterns of transit of data and ideas relating to humankind's deep past. These sections analyze the tumultuous subcontinental geography of theories of human development in the wake of Calcutta's establishment as a hub in world-spanning networks of knowledge. To a far greater extent than anthropologist-historian Bernard Cohn acknowledged, the colonial "command of language" and associated hypotheses of humanity's origins in South Asia were both fractured and fractious.[5] From the 1810s through to the mid-nineteenth century, there occurred a series of challenges to the supremacy of Jones's theory of a unitary Indian branch of the linguistic-racial "tree" and, simultaneously, to Calcutta's hegemony as a center of knowledge production. These contestations emerged first in Madras and then in Bombay—rivaling Calcutta as seats of British colonial power—and formed the beginning of theories of a bifurcated or trifurcated population of India. Further interventions sought to displace these coastal Presidency cities and promote interior uplands and mountainous and desert-bound fringes of the subcontinent to the status of true centers of the deep past of humanity. India's geographies of knowledge of humanity's deep past were marked during the nineteenth century by multiple decenterings to previously unheralded locales rather than by the production of increasingly consolidated epistemic hubs. The chapter shows that during the nineteenth century areas peripheral to the colonial state's commercial, fiscal, and governmental interests came to be deemed integral to human diversity studies.

The "people of India"—the would-be objects of analysis for Risley and his many European precursors—were themselves active players in the game of theorizing humankind's deep past. In the chapter's third and fourth sections I demonstrate how inhabitants of the Indian subcontinent were much more than mere informants for European theorists. Reforming urbanites, religious scholars, and inhabitants of fringe upland and "jungle" areas all questioned the origins and migrations of humans and considered the contemporary and future implications of these conjectures about the past. Though often inflected and provoked by Western theories, these debates were variegated and never straightforwardly derivative.[6] The supposedly bulldozing concepts of European science relating to the human past—most notably Darwinian evolutionary theory—entered into some of South Asia's intellectual milieu, but in doing so the concepts changed profoundly. Like their European contemporaries, Indians often entwined evolution with religious narratives and agendas; and in common with the British evolutionary writers considered by Ian Hesketh (this volume), many saw historicizing humans as an undertaking that envisioned futures as well as reconstructing pasts. In the intellectual sphere, as well as the political, the age of high empire was an era of burgeoning resistance and renewal among many colonized societies, and nowhere more so than in

India. While colonial domination lent European knowledge an allure, this was rarely a cue to accept its tenets without significant reworkings that employed elements of the subcontinent's deep ideational heritages. In undertaking such selective appropriations, Indian thinkers turned the tables on many of the European progenitors of the ideas they used and altered, who had relied on recasting the thoughts of informants in colonial locales in various innovative—and disingenuous—ways.

This chapter builds on work that has overturned the notion that scientific theories and practices diffused from Europe to the rest of the world during the age of empire.[7] The ideas developed by India's inhabitants were not simply "local" in comparison to the "global" reach of the foremost theories of human origins developed by the colonizers. The globality of European science functioned only at the level of its self-proclaimed explanatory power, and many Indian theories of human development also claimed universality.[8] Debates in the Indian subcontinent inspired some thinkers—both Indian and European—to go further still, subsuming the science of humanity's past into cosmological concerns about the nature of matter, the universe, and the divine.

INDIA AS THE SPACE OF HUMAN ANTIQUITY

Two men sit on adjacent sides at the corner of a desk, which is scattered with ink-pots, pens, and papers written in various alphabets. Among the volumes is a book that one of the men has compiled, and the other is struggling to understand. He is struggling not because he cannot decipher the script or because of the incomprehensibility of particular words and sentences. It is because at a fundamental level he does not agree with the other man and his text.[9] He thinks that what they convey is wrong—and more than wrong, "absurd" and "wholly fabulous."[10] So when he asks the other man for clarification, he does so hoping he will find it useful, even provocative, but he knows he will not believe it.

This scene took place in Calcutta in late 1787; the book at its heart was a compendium of ancient Hindu texts, the Puranas; its compiler and interpreter was a Bengali Brahman (high caste Hindu), Radhakanta; and the skeptical interlocutor was a forty-one-year-old London-born judge and polymath, William Jones. Jones was fascinated by the very aspect of the Puranas that he also scoffed at: their account of the origins of the world and of humans. Since his arrival in Calcutta four years previously—and when he had time away from the courtroom—Jones focused on what he claimed was "the great problem": whether the various "nations" of people across the world "had *any* common origin, and whether that origin was *the same*, which we generally ascribe to them."[11] Jones's answers to his self-described "problem" became profoundly influential, setting an agenda for major studies of human origins during the nineteenth century in colonial India and metropolitan Europe. Before investigating their dispersal and impact, it is necessary to understand the knotty interactions through which Jones's theories developed.

Radhakanta believed the truth of humanity's origins was related in Sanskrit scripture, which outlined a cyclical view of human development stretching back nearly two thousand million years. Jones, while admiring and thoroughly reliant upon Radhakanta's learning, told his fellow members of the Asiatick Society, the institution he had founded in Calcutta within months of his arrival, that this timescale was "repugnant to the course of nature and to human reason."[12] He blamed Radhakanta's community for disingenuously promoting this lengthy human history, terming it "fictions of the Brahmans" peddled with a view "to aggrandize themselves."[13] Jones's suspicious engagement with Brahmanic knowledge was instead geared toward calibrating it with a biblical timescale of approximately six thousand years. The Hindu chronology, he hypothesized, may be "the same [as] our own, but embellished and obscured by the fancy of their poets and the riddles of their astronomers."[14] He placed particular significance on the existence of a flood narrative in the Puranas, which he claimed constituted independent verification of the truth of the biblical deluge.[15]

India and Indians served a much more productive role in Jones's ethnology than merely being sources of error to be criticized and disciplined into conformity, however. His use of Indian sources and informants crucially shaped his contentions: India changed Jones as much as he interfered in India's learned traditions. His scholarship was a central component of the vibrant and complex intellectual milieu in Calcutta during the 1780s, which, as Kapil Raj has shown, was a key "contact zone" in which cross-cultural epistemic networks flourished.[16] Jones also conformed to long-running European notions that

India was the cradle of humanity, the key to understanding the emergence of civilization. In accounts from the seventeenth century onward, and most famously in Voltaire's polygenism of the eighteenth century, India's venerable civilization shook up Christian norms and biblically derived theories.[17]

Jones's claim that ancient Hindu texts were vital to understanding human development on a global scale was in keeping with an already established trend. But his use of Brahmanic scholarship and the richness of his theoretical schema far exceeded his predecessors' work. Jones placed Sanskrit at the core of both his method and his conclusions. His postulation of a link between Indians and Europeans (not to mention various other groups, including "the old Persians, Ethiopians, and Egyptians, . . . the Chinese, Japanese, and Peruvians") and his claim that the cradle of postdiluvian humanity lay in Persia rested on the Sanskrit language itself.[18] Jones radically diverged from his predecessor Alexander Dow (1735/6–1779) in suggesting that Sanskrit was not "beyond" other languages but, rather, "bear[s] to both [Greek and Latin] a stronger affinity, than could possibly have been produced by accident." The three must, he claimed, have "sprung from some common source."[19] He went on to extend his analysis of grammatical structure—distinct, he was at pains to insist, from the "fallacious" etymological method employed by previous British theorists such as Jacob Bryant—to encompass the ethnology of all Eurasia and some areas of the world beyond.[20]

The object of his major intellectual endeavor in India was to shore up the foundations of the biblical account of postdiluvian human development against prominent Francophone enlightenment thinkers such as Voltaire (1694–1778).[21] But Jones's theory was put to much broader uses in the nineteenth century. Along with other Calcutta orientalists' theories and methods, it spread through epistolary networks and print cultures into not only metropolitan Europe but also the Malay Archipelago and the South Pacific.[22] This dispersal of texts (and, increasingly, materials) meant that India lay at the center of some of the leading studies of human origins around the turn of the nineteenth century. The enormously influential *Researches into the Physical History of Man* (1813), by James Cowles Prichard (1786–1848), drew extensively on the Asiatick Society's widely circulated journal, *Asiatick Researches*. Prichard echoed Jones in all key respects, from the language-based methodology to his deference to Vedic texts, to positing an arc connecting India, Persia, and

Egypt as the cradle of civilization.[23] Prichard also drew on John Bentley's work in *Asiatick Researches* on Indian astronomy to fit "the Hindus" into a biblical chronology, while disparaging the "extravagant pretensions to antiquity" of the Vedas' vastly extended conception of the human past.[24] Prichard's revised and hugely expanded third edition of his *Researches* (published 1836–1847) privileged a wider range of data and theories from the Indian subcontinent, as well as repeating with even more vigor his critique of the "fabulist" Vedic timescale.[25]

Jones's theories on the origins and varieties of humans fed into diverse intellectual spheres, in which linguistic, climatic, and physical methodologies jostled for priority.[26] Similarly, his suggestion of shared postdiluvian origins of Indians and Europeans and hints at the temporal priority of Indian civilization did not have a straightforward single implication for colonial governance but, instead, influenced acrimonious discussions on the subject throughout the nineteenth century. The concept of common Indo-European origins disrupted any easy notion of the colonizers' inherent or long-running superiority over the colonized. As Thomas Trautmann has shown, this notion of the identity of rulers and ruled continued to provide the underpinnings for a more inclusive form of colonialism into the later nineteenth century, with Friedrich Max Müller's (1823–1900) theory of Aryanism at the forefront.[27]

Temporally distant kinship did not, however, generate a robust conviction of the equality of contemporary European and Indian people and society. India's antiquity was used frequently as a means through which British figures suggested the degeneracy of its present—or at least its temporal dislocation from modern Europe. Jones told a wholly European audience in 1786, "We now live among the adorers of those very deities, who were worshipped under different names in old Greece and Italy, and among the professors of those philosophical tenets, which the Ionick and Attick writers illustrated."[28] Such pronouncements, which British thinkers including Henry James Sumner Maine (1822–1888) echoed throughout the nineteenth century, not only relegated contemporary Indian society to a survival of the distant past but also "show[ed] how the new colonizer was in some sense already present in the colonized territory."[29] Kinship could be taken to imply the legitimacy of British rule over their supposedly degraded Aryan brethren. Common heritage in a distant past provided an "alibi of empire" at least as often as it gave reason to doubt British interference in India.[30] The governmental implications of Jones's theories were highly contested and never stably resolved. The same could be said of his impact on notions of human origins and variety within and beyond colonial India. What was indisputable, however, was the spatially dispersed influence of Jones's placing India at the core of his rendering of humanity's deep past.

DECENTERING THE SPACES OF SCIENCE

William Jones admitted no doubt over the key conclusions of his researches in India: the validation of "the Mosaick history" and the descent of Europeans, Indians, and Egyptians from Ham.[31] Nonetheless, the edges of his theories were frayed, leaving numerous questions unanswered and data unclassified. Especially vexing difficulties presented themselves when Jones attempted to account for the diversity of people beyond the fringes of British governmental control in the Indian subcontinent at that time. He suggested that the people of Sind and the inhabitants of the uplands of northeastern India were of Hamitic origin but were separated early from the other people of India and, in the latter case, "soon intermixed with the first ramblers from Tartary."[32] Jones even admitted a mysterious element at the core of his theory, identifying a small portion of Hindi—the prevalent language in north India—that derived from a language other than Sanskrit. Although Jones's major successor in Calcutta, Henry Colebrooke (1765–1837), claimed in the first decade of the nineteenth century that Hindi derived in full from Sanskrit, this contention came under substantial scrutiny in ways that intermingled philology and studies of human variety.[33] The challenges to the Calcutta orientalists' account of human origins that emerged during the early to mid-nineteenth century were also struggles over intellectual and institutional authority from locations in colonial India beyond Bengal. These debates foregrounded regions that had previously been framed as peripheral, the relative isolation of which supposedly confirmed their status as repositories of humanity's deep past.

Madras had long formed a base of British political, economic, and epistemic power in India, distinct from—and often in opposition to—Calcutta. It was the location of a major astronomical observatory and the cradle of British trigonometrical surveying in India. Its scholar administrators led the way in aspects of colonial medicine and botany.[34] Madras was also where, in 1816, Alexander Campbell (1789–1857) published the *Grammar of the Teloogoo Language*, a work that, despite its uncontroversial title and limited intended circulation among students at the College of Fort St. George in Madras, contained an incendiary central thesis that influenced a far wider audience. Campbell drew on the grammatical analysis of Francis Whyte Ellis (1777–1819), a Madras civil servant of twenty years' experience, and on his own extensive interactions with southern Indian pandits. Placing these authorities in direct opposition to the Calcutta orientalists, Campbell claimed that categories drawn from Madras's grammarians indicated that southern Indian languages, including Telugu, Tamil, and Kannada, were not derived from Sanskrit. Instead, as Ellis put it, they formed "a distinct family of languages, with which the Sanscrit [*sic*] has, in latter times especially, intermixed, but with which it has no radical connection." Campbell and Ellis's analysis also insinuated that the population of their region was quite separate from, and inhabited India prior to, the speakers of Sanskritic languages to the north.[35]

Robert Caldwell (1814–1891), a missionary based in Madras Presidency, further developed this thesis in an influential 1856 work, innovating the term "Dravidian" to cover southern Indian speakers of non-Sanskritic languages. Caldwell stated bluntly that "the arrival of the Dravidians in India was undoubtedly anterior to the arrival of the Aryans," and he also posited a separate, pre-Aryan group in northern India.[36] The work of these Madras orientalists over four decades or so constituted a major redefinition of the Calcutta orientalists' account of human and civilizational origins and variety in India. As Thomas Trautmann has suggested, it sought to subsume Calcutta's intellectual authority into that of Madras, proclaiming that the latter region had the informants and methodological insights to develop an accurate account of human variety throughout the Indian subcontinent.[37]

By the time of Caldwell's major publication, other versions of the notion that Sanskrit speakers were not the earliest inhabitants of the Indian subcontinent had arisen beyond Madras. Alexander Hamilton (1757–1804), the first professor of Sanskrit in Britain and a major backer of the Calcutta orientalists, had mooted in 1808 that "the wild but harmless inhabitants of the mountains" in India might be "a race distinct from the Hindus." This suggestion garnered attention from a number of institutions and agents operating from the intellectual fringes of British India during the mid-nineteenth century.[38] Though an arena of long-established colonial commercial power, Bombay lacked the intellectual prestige of the other Presidency cities of British India, Calcutta and Madras. This began to change with the investigations into the languages and peoples of the hinterlands around Bombay undertaken by John Stevenson (1798–1858), beginning with his 1841 text, "Observations on the Marathi Language."[39] Stevenson, like Caldwell a missionary influenced by Campbell's work on Telugu, couched Bombay—and the city's branch of the Royal Asiatic Society, of which he was president—as the crucial location for studying the original distinction between Sanskritic and pre-Sanskritic languages. Marathi, he claimed, marked the border between the two. Assuming the identity of language and race in line with his contemporaries in British India, Stevenson argued that Marathi afforded glimpses of the "rude aboriginal population" that inhabited all India before the Aryan invasion drove many southward and into upland regions and incorporated some others as the lowest caste.[40]

This story of Aryan invasion and the removal of the original inhabitants of India held an appeal for many agents of the Company-State during its last great era of territorial expansion, another instance in which human diversity studies provided an alibi for empire. The story also engendered an emphasis on regions at the outskirts of colonial governmental control, in attempts to comprehend the variety of humankind. One gazetteer was typical in presenting the hills of central India as a "great natural fastness" for the aboriginal tribes.[41] Studies of humanity's deep past that centered on those Indian spaces couched as isolated and primeval also tended to mingle—in conceptual and material terms—geological strata, fossils, animals, and people.[42] The shift of focus to India's uplands and the overlap of earth and human sciences were both elements in the work of Brian Hodgson (1800/1–1894), who, along with Stevenson, was a progenitor of an "aboriginal" theory of humanity in India. Hodgson's theories of human diversity formed part of his prodigious scholarly output, which also included botanical collaboration with Joseph Dalton Hooker (1817–1911), following his removal from residency at the Nepalese Court in Kathmandu in 1843.[43] His model of human development in South Asia shared notable similarities with Stevenson's, including the concept of a single aboriginal population of South Asia that was shattered by Aryan invasion.[44] Its key difference was one of location: Hodgson's removed location—

in the Himalayan town of Darjeeling at the northern fringes of India—and immersion among upland communities informed his distinctive conception that the Aryans had exercised "savage tyranny" over a previously thriving aboriginal civilization.[45] Hodgson's sympathy for inhabitants of hill and mountain regions and his distaste for attempts to "Anglicize" India, in vogue among administrators in Calcutta, underlay his contention that the invading latecomers, not the aborigines, were the real barbarians.[46]

Hodgson's account of human origins in India might be said to encompass elements of what Thomas Trautmann terms "Indophilia" and "Indophobia" in British attitudes to the Indian subcontinent's inhabitants. Hodgson's long double career in the administrative and intellectual realms spanned a period of increasing disparagement among British administrators and scholars toward Sanskritic Hindu culture. Although far from the violent extremes of some of these views, Hodgson's anti-Aryanism was substantially compatible with them. His pro-aboriginal thought was also broadly compatible with the growing valorization of "primitive" communities in upland and forested regions, especially in northeastern India, among some colonial administrator ethnographers.[47] Although the convictions Hodgson shared with Stevenson and Ellis—that language and physiology told the same stories of human difference and that India's aborigines were an originally unitary group—were called into question during the 1850s by Caldwell and others, his theories intertwined with longer-lasting trends within colonial thinking about the peoples of India.[48]

These suggestions of the heterogeneity of India's population greatly impacted metropolitan studies of human diversity during the crucial mid-nineteenth-century contests over a distinct discipline of ethnology. Among the most notable influences of these studies was on Prichard's vastly expanded third edition, complete with its subtly altered title, of *Researches into the Physical History of Mankind*. Prichard greatly increased the space given over to discussing India in this edition compared to his 1813 original, in significant part to accommodate the enormous influx of ethnographic information and ethnological theorizing from India's other Presidencies and upland margins. He argued that Colebrooke "appears to have adopted . . . from Sanskrit authors' the misguided notion of the Sanskritic origins of southern Indian languages. Instead, Ellis's 'more accurate research' in Madras 'fully prove[s] . . . a distinct family of languages, which . . . have in reality no essential connexion with [Sanskrit]."[49] Prichard followed Stevenson in hypothesizing aboriginal unity in India, writing that "the civilized Dravirians [*sic*] . . . descended from the stock of the wild aborigines" were most probably Tartars who had entered through northeastern India.[50]

Prichard also believed that further information on the people of the Subcontinent, produced in the decades between the first and third editions of his magnum opus, constituted prime evidence to refute increasingly influential theories suggesting physical fixity of racial groups. He argued that members of "the Indo-European or Arian race" varied greatly in their skin tones, which backed up his key contention of the importance of climate in inducing physical variety. India provided especially compelling evidence in this respect because, Prichard suggested, the strictures of the caste system ensured that color variations were not the result of "intermixture" between discrete groups (a contention Risley echoed half a century later).[51] One of the major pieces of evidence that Prichard marshaled for his climatic theory was the "sanguine or xanthous complexion [akin to] the northern European" of the Kalash people in the extreme western Himalayas, which he attributed to them "being undoubtedly of the Hindu race, which their language attests, and inhabiting a high and cold country."[52] Focusing on an upland community was wholly in keeping with the increased significance of India's peripheries in studies of human variety from the 1810s onward, and this also occurred among others who conveyed information from the colonial Indian subcontinent to metropolitan audiences. In 1844 Thomas Postans (1808–1846), a mid-ranking ex-soldier administrator who spent the later years of his career on the distant desert- and hill-bound frontier of Sind in the extreme northwest of India, gave one of the earliest papers at the recently founded Ethnological Society of London (1843). Postans proposed that the seemingly anomalous Baluchi community he encountered in this area "form[ed] a connecting link . . . between the Persian and Affghan [*sic*] tribes beyond, and the mixed Rajput races" and was therefore vital to studies of human diversity.[53]

Indian locales that were peripheral to colonial government and commerce came to be seen as central to human diversity studies during the first half of the nineteenth century. These regions were credited by metropolitan authorities such as Prichard, as well as Europeans in India, as being repositories of pure aboriginality in the Indian subcontinent. They were seen to hold the key to understanding the population of this supposedly antique land. Intermingled with this focus on the fringes was a dynamic set of contests between rival centers of knowledge production in colonial India. Tussles over who and where had a privileged claim to provide an accurate account of human variation formed a significant element in efforts first in Madras and then in Bombay to assert epistemic supremacy over Calcutta. In terms of knowledge, as well as profiteering and politics, the Indian subcontinent was a place of contestation between colonial officials and institutions, and beyond them too.

NEUFVILLE AND THE SINGPHOS

India's fringes were not merely arenas of "information famines"; they were integral to various scientific endeavors.[54] The following brief case study relates this insight to an issue discussed briefly with reference to William Jones's scholarship: the place of indigenous agency in colonial studies of human diversity. Whereas Radhakanta's and Jones's other informants were members of already established learned elites, the example considered here indicates dispersed agency in theories of the origins of humankind. British contributors to debates on humanity's deep history, who were themselves often relatively marginal within colonial knowledge structures, could credit people from the edges of empire with epistemic authority. But this authority tended to be mediated and circumscribed—restrictions that various Indian thinkers fought against.

Let us turn to the eastern limits of the Brahmaputra valley in the 1820s, as British East India Company forces advanced well beyond the Company's territory in Bengal into regions with which it hitherto had very few dealings. Among the ranks of this invading army were officers with more than a military interest in the area. These officers' attempts to know the land and know the people blurred into a unified endeavor, exemplified in an 1828 article by John Neufville (1795–1830), a soldier official about whom only scanty fragments survive in the colonial archive.[55] Mixing hydrography of the expansive Brahmaputra River system with ethnography of the equally complex and numerous communities in the region, Neufville's writing indicated the ambivalence of British administrator scholars' engagements with local informants, in which reliance and suspicion were side by side.

The accessibility of informants and the administrative concern to gather knowledge of the region's most troublesome communities dictated the focus of Neufville's ethnography. At the intersection of both elements were the Singphos, a group that straddled the lowlands and hills at the extreme eastern edge of the Brahmaputra valley. Neufville's account of the Singphos started with an irate denunciation of the "utter havoc and desolation" they wrought, as they descended into the Brahmaputra valley, but then the account transitioned into a more sensitive mode.[56] In "endeavoring to trace their manners, customs, and traditions," Neufville placed particular emphasis on the Singphos' accounts of their creation and dispersal. His rendering of these theories stated that the original Singphos were created atop a mountain located in present-day northern Myanmar and thereafter washed downstream toward the Irrawaddy River. For a period they halted and "were immortal, and held celestial intercourse with the planets and all heavenly intelligences, following the pure worship of the one supreme being," but upon descending to the plains, twenty-one generations prior to Neufville's day, "they fell in with the common lot of humanity, and . . . soon adopted the idolatries and superstitions of the nations around them."[57] It is at the point of the Singphos' supposed descent into the valley lowlands that Neufville's interests began more clearly to guide the narrative. A history of Singpho labor systems and authority structures took center stage, guided by administrators' attribution to the Singphos of morally repugnant practice of slaveholding and the colonial state's need to identify powerholders within the community.[58]

The preceding narrative of Singpho origins and migrations, however, revealed tensions that ran through much British discourse concerning Indians' theories of human origins. Neufville was never quite sure how much credence to give to the Singphos' stories. On the one hand, he quoted at length a testimony largely derived from a Singpho potentate in an article published in *Asiatick Researches*, which conferred upon it a degree of scholarly authority. On the other hand, Neufville's interventions in and skepticism toward this narrative are both apparent. He chose this account from a number given by various Singpho sources as "it appears to be the most consistent" and the informant was "the most intelligent of them." Nevertheless, Neufville cross-referenced elements of this tale with the others and excised its "fabulous portions," confining the supposedly unabridged version to an appendix.[59] His separation of the "fabulous" rendered Neufville the author of a distinct version of the Singpho creation story, an act that simultaneously positioned himself as the proper arbiter of a reasonable theory and suggested that Singpho narratives contained enough credibility to be a relevant element in establishing how the community came to occupy its current geographical and societal position.

The two Singpho creation accounts, included in the appendix to the article, claimed to have explanatory power for all humanity, invoking (in Neufville's written translation, of course) "mankind" and "the race of man" respectively.[60] Neufville's omission of such universal categories in his version of the Singpho creation myth—instead saying merely "the *Sinh-phos* were originally created"—sought to downscale and localize the theory's explanatory power. However, the fact that he did not subsume the Singphos into any meta-scheme of human origins and migrations gave his Singpho-derived narrative space in which to appear credible. He did not remove the particularities of this theory through placing the Singphos within a biblical framework with global pretensions, nor did he invoke the authority of philological or physical analyses to destabilize the Singphos' own accounts. These features of Neufville's article stand out still more when read alongside the influential account of the Singphos published forty-four years later by the administrator ethnographer Edward Tuite Dalton (1815–1880), which claimed they first entered the Brahmaputra valley from Burma in 1793 and had "features . . . of the Mongolian type."[61] Despite Neufville's flashes of skepticism and his self-assigned editorial control over the Singphos' creation story, the account in *Asiatick Researches* provides us with one example of how colonized and semicolonized communities—even ones without a written language—possessed nuanced notions of creation and historic migration. These theories had complex relations to European science: they were not straightforwardly local mythologies counterposed to global knowledge, and British attempts to render them such were unstable. Thus, Indian data and theories of human development had the ability to "disturb" European knowledge.

INDIAN INFLECTIONS AND THE DISTURBANCE OF EUROPEAN KNOWLEDGE

Information and materials from India were put to use in diverse ways in European and North American accounts of human origins and diversity. These deployments were prone to slippages and were contested by rival theorists and members of the populations they sought to analyze. Before shifting its focus to reconfigurations of concepts of evolution among Indian intellectuals, this section opens by examining an instance of the mutability of Indian data in metropolitan debates: the various depictions and discourses relating to human diversity studies that invoked the Bengali intellectual and reformer Rammohan Roy (1772–1833), who was possibly the most discussed subject of empire in early nineteenth-century Britain.

Although it is uncertain whether Prichard met Rammohan on the latter's visit to Britain in the early 1830s, Rammohan's portrait appeared a little over a decade later as the frontispiece to a volume of *Researches into the Physical History of Mankind* (fig. 5.1).[62] Prichard wrote that the image "displays the countenance of a very dark Brahman," and evidenced his claim that Rammohan "was much darker than many Africans."[63] Rammohan's complexion, Prichard avowed, was indicative of the variety of skin tone within each caste, which in turn exemplified a core element of his theory of human diversity, namely, "the very real process of deviation, according to which varieties are produced in human races as in the inferior orders."[64] These textual and pictorial renderings of Rammohan were dubious. The frontispiece image was derived from an 1822 engraving produced in Calcutta by Phillippe Savignhac, a Frenchman, a fellow Unitarian of Rammohan's, and later an official draftsman and engraver to the Calcutta Phrenological Society. The original engraving and its reproductions in various Unitarian publications in 1823–1824 were monochrome and did not firmly suggest the coloring in Prichard's portrait.[65] A number of other portraits produced around the time of Rammohan's residence in Britain, including an anonymous watercolor rendering of Savignhac's engraving, took a radically different approach to his skin tone, portraying him as much fairer than Prichard's frontispiece.[66] The coloring of the frontispiece in *Researches into the Physical History of Mankind* was, to say the least, convenient for Prichard's thesis, an example of the tenuous rework-

FIGURE 5.1. Portrait of Rammohan. From James Cowles Prichard, *Researches into the Physical History of Mankind*, vol. 3. (London: Sherwood, Gilbert, and Piper, 1841), frontispiece. Reproduced by permission of the Master and Fellows of Gonville and Caius College, Cambridge.

ing of Indian data at a temporal and geographical distance to fit a particular account of human variation.

Rammohan's varied figurations in competing theories of human diversity further contested Prichard's portrait and description. Following his death in 1833, one of men who tended to Rammohan in his final hours sent to the Edinburgh Phrenological Society, the leading institution in its field, a cast of Rammohan's skull, taken, he avowed, "when the body was then quite warm."[67] The analysis was written up along with a potted biography in the society's journal and directly contradicted Prichard's climatic theory of human diversity. The article opened with a diatribe against ascriptions of physical variation to the influence of climate, stating that this outlook "is inconsistent with observed and established facts."[68] It featured an image of Rammohan very different from Prichard's in order to "convey to the reader an accurate general idea of the head" (and to suppress potential errors in the casting process) in support of its anticlimatic contention (fig. 5.2).[69] Different renderings of the same evidence—in this instance, a single element of Rammohan's anatomy—could be integral elements within diametrically opposed analyses of human diversity. At a time of fierce debates over climatic influence and physiological fixity in human physiology and moral intellectual capacity, people, materials, and information from the Indian subcontinent were harnessed into far-flung but fragile networks.[70] They were taken to be vital sources but, as in the case of Rammohan's head, produced contestable, unsettled knowledge.

Epistemic mutability could go further still. In debates on the diversity of humanity, Rammohan was not simply a distended head, variously represented by metropolitan theorists in support of competing ideas. He was an active agent, collecting skulls of supposedly typical "Hindoos" for Edinburgh phrenologists during the 1820s, before later (unlike many fellow educated Bengalis)

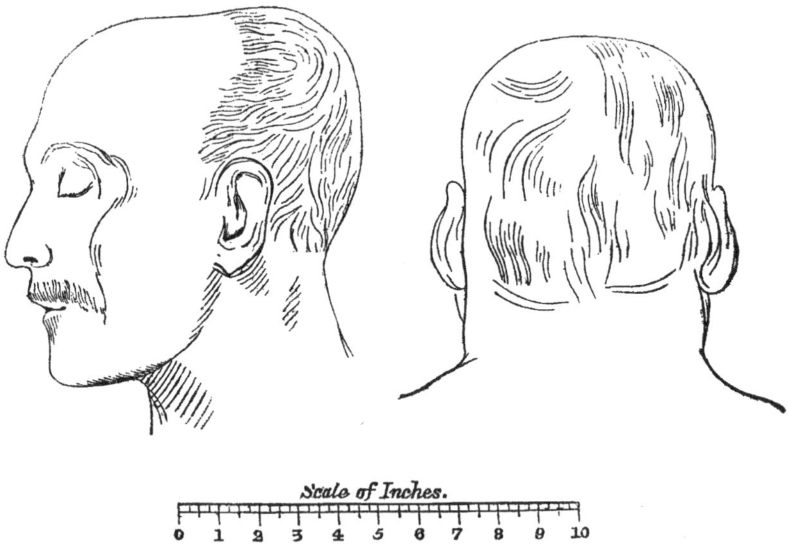

FIGURE 5.2. Image of Rammohan's head. From Anonymous, "On the Life, Character, and Cerebral Development of Rajah Rammohun Roy," *Phrenological Journal and Miscellany* 8 (1834): 579. Reproduced by permission of the Whipple Library, University of Cambridge.

dismissing phrenology.[71] While only in rare instances as mobile or influential as Rammohan, many of the Indian subcontinent's inhabitants were active participants in making knowledge about the human deep past. Even the advent of the Mosaic explanation of human origins in the Indian subcontinent was not the preserve of Jones and the British but featured in Mughal-era histories by Muslim intellectuals, which categorized Hindus as descendants of Ham.[72] In the decades following Rammohan's death, Indians engaged racial theories premised on language and on physique. Among the most notable instances of the latter was the foundation in 1845 of the Calcutta Phrenological Society by Kali Kumar Das, part of the mid-nineteenth-century phase of the "Bengali Renaissance."[73]

Let us now turn to the key Indian theories of human development in the later nineteenth century, an era where variegated notions of evolution took center stage and debates on human diversity intensified. Any notion that the Rebellion of 1857–1858 and the more or less contemporaneous publication of Darwin's *On the Origin of Species* led to abrupt changes in discussions of human diversity in India is fallacious.[74] These elements instead fit into subtle, longer-term patterns of political and intellectual continuity and change. Race had mattered in colonial India long before the Rebellion, and the supercharged debates in metropolitan ethnology during the 1860s, and the assumed analytical purchase of racial categories remained deeply contested during the following decades.[75] Darwinian ideas also percolated unevenly and, in many cases, far from immediately into the Indian subcontinent's rapidly developing public spheres. Furthermore, Darwinism had competition from alternative understandings of human development, including Friedrich Max Müller's Aryan theory. Although the onslaught of physical anthropology in Britain during the 1860s led Müller to draw back from his earlier claims on the equation of language and race, his previous message of brotherhood between Europeans and Indians unleashed what Tapan Raychaudhuri termed, "a spate of 'Aryanism'" among portions of the educated elite in India, which lasted into the later nineteenth century.[76]

The core concerns in contemporaneous British ethnology did not dictate Indian elites' political and theoretical engagements with ideas of human diversity from the 1860s onward. Nevertheless, evolutionary theories came to prominence in learned circles in India, especially in Bengal, during the crucial period in the late nineteenth century when Indian national politics and religious revivalism were fermenting. Those engaged in present-minded debates that often invoked prescriptive demands for the future adapted ideas of the past from European thinkers because these often seemed prestigious and useful. They melded these ideas and frameworks with other modes of thought that were already extant or in the making and created ideational products that elided the categories of local and global, traditional and modern. As Marwa Elshakry has commented on the Arabic-speaking world, the key to the spread of Darwinian evolution (and of new notions of humankind's past, more broadly) was its ability to be assimilated, its "brilliantly ambiguous" quality.[77] The dynamic ideational milieu of India produced theories of human origins and variation that were, in some instances, influential on scales comparable to that of the spread of Darwin's ideas.

The truism that ideas morph in transmission was evident in specific ways in Bengali elites' discussions of evolution. Like their contemporaries in Britain, some Indian thinkers applied Darwinian ideas to humans before Darwin himself did in *Descent of Man* (1871).[78] While Mahendralal Sarkar (1833–1904), later founder of the Indian Association for the Cultivation of Science, declared in an 1869 lecture that "the Darwinian hypothesis of progressive development . . . is daily receiving confirmation," other intellectuals ridiculed Darwin as "a joke," focusing their amusement on the concept of ape ancestry.[79] The variety of Indian opinions regarding human applications of Darwin's thought during the late 1860s and early 1870s indicate the problem with any uniform reading of Darwinism's impact in the Subcontinent, whether Pratik Chakrabarti's claim that it was "widely contested" or Deepak Kumar's suggestion that it was "imported readily."[80]

Indian deployments of Darwinian theories of human development were radically diverse. In India as in the Arabic-speaking world, and also in Britain, Europe, and North America, it is clear that Darwin was often "read through other readers," and that a range of ideas was ascribed to him—disingenuously or otherwise—beyond those detailed in his major publications.[81] Indians confronted not a unitary or pure Darwinism but many Darwinisms; and their own acts of interpreting and devising new theories generated still more multiplicity. The lexicon of evolution reached many intellectuals not through Darwin alone but partly or primarily through the works of Herbert Spencer

(1820–1903) and others.[82] Keshab Chandra Sen (1838–1894), a syncretic thinker who drew on Christian and Hindu concepts as well as evolutionary theories and was a close associate of Müller, spoke collectively of "the Darwins and Huxleys, the Tyndalls and Spencers" and elsewhere explicitly set aside Huxley and Darwin and their discussions of "protoplasm [and] natural selection."[83]

We should also treat with caution attempts to invest explanatory power for the uptake of evolutionism in apparent points of compatibility between Hindu cosmology and Darwinian accounts of human development. Two of these supposed commonalities were the evolution of humans from animals and the projection of a long human chronology. As C. Mackenzie Brown has shown, both of these elements also contained pertinent points of difference between mainstream Darwinism and traditional Hindu texts, which emphasized the exceptional status of humans, implying fixed rather than evolving species, and put human (and animal) origins nearly two thousand million years in the past, a timescale that dwarfed Western estimates.[84] According to some influential Indian thinkers, the Vedas' precise enumeration of human antiquity went sharply against Darwinism. The most notable of these thinkers was Dayananda Saraswati (1824–1883), the founder of the Hindu revivalist organization the Arya Samaj in 1875. Dayananda insisted on the literal truth of Vedic texts, which by his reading placed humanity's origins on Earth in Tibet and suggested that humans also existed on all celestial bodies, including the sun and moon. He found the concept of ape ancestry among "the followers of Darwin and philosophers of his school" especially problematic and asked, in a public lecture, "If man was really a descendent of a monkey . . . how was it that for thousands of years past, no monkey's young one had developed into a human being?"[85] Rather than seeing Darwinism and Hinduism as inherently compatible and using this insight to explain the uptake of the former in India, we should focus on what was at stake when Indian thinkers suggested that these two frameworks shared common ground.

Though distinct in their thought and politics, the Bengali intellectuals Bankim Chandra Chattopadhyay (1838–1894), and Swami Vivekananda (1863–1902) were among those who deployed Darwinian theory as part of broader attempts to subsume the authority of European science into Hindu cosmologies. Bankim posited that natural selection was a power complementary to the creative and preserving powers overseeing the evolutionary development of species, making Darwinism an adjunct to the Hindu trinity of Brahma, Vishnu, and Shiva.[86] Through this logic Bankim claimed that "in comparison with Christianity, the religion followed by the great practitioners of science, the European peoples, the Hindu worship of the Trinity is far more natural and in accordance with scientific theories."[87] Vivekananda, whose conception of evolution during the 1890s drew in Lamarckian and Spencerian theories and downplayed the Darwinian notion of the struggle for existence that appealed to many British theorists of humankind, went further than Bankim's trinitarian reading of evolution.[88] Vivekananda concurred with the first Boden professor of Sanskrit at the University of Oxford, Monier Monier-Williams (1819–1899), who suggested, "The Hindus were . . . Darwinians centuries before the birth of Darwin, and evolutionists centuries before the doctrine of evolution had been accepted by the Huxleys of our time."[89] Claiming that "the theory of evolution . . . is the foundation of almost all the Indian schools of thought," Vivekananda espoused what we might think of as a reverse diffusion model, with evolution making its way from the Vedas to "the physical science of Europe."[90] He and Bankim were among many others who attempted to craft distinctive modernities that put ancient Sanskrit texts, Bengal, and India at the center of intellectual frameworks with universal ambitions. When seen in this light, their engagement with theories of human origins and development—and the idea that the progenitors of these theories had a privileged claim to modernity and civilization—shared notable similarities with the Calcutta orientalists' schemes around a century before. For them too, humanity's deep past was a "great problem."

It was not, however, *the* great problem it had been for William Jones. Most Indian thinkers who had engaged with Darwinism and theories of evolution did so in ways that reduced the emphases on the means by which, and time and location in which, human beings first appeared, as well as the details of their subsequent development. One especially striking instance of this was Keshab Chandra Sen's claim that "the question is perhaps not so serious after all, whether men have descended from inferior animals. . . . [But] in the individual there is something like evolution going on unceasingly."[91] This evolutionary process entailed freeing oneself from "animality"—a lingering remnant of "our peculiar ancestry"—and progressing to "a yet higher stage of development . . . Godliness."[92] Keshab's future-oriented and spiritually focused deployment of evolution was influential, informing major figures including Vivekananda. By this reckoning, physical evolution and a focus on past developments were utterly insignificant compared with spiritual evolution and progress toward future cosmic unity. The leading Bengali intellectuals as the twentieth century dawned, Vivekananda and Aurobindo Ghose (1872–1950) interrogated humanity's potential development into the divine, instead of its historical development from animals. Moreover, they critiqued materialistic and mechanistic accounts of evolution as leading to the privileging of immediate goals and competition at the cost of ignoring higher goals of perfection, as outlined in the Vedas.[93]

Aurobindo may have exaggerated in claiming that "the materialistic view of the world is now rapidly collapsing and with it the materialistic statement of the evolution theory must disappear," but the spiritualized version of development that he and others propounded spread far beyond Bengal and South Asia.[94] For instance, the Theosophist movement, which garnered significant support in Europe and North America, placed significant weight on antimaterialist evolutionism.[95] As the twentieth century dawned, India remained a vital hub in world-spanning networks of theories that historicized humans. This influence was, however, centered on thinkers who insisted upon the inadequacy of what they understood as Western science as much as it centered on British colonialist-anthropologists such as Herbert Risley who worked within this tradition. In India, European understandings of human diversity and development did not diffuse; they were instead appropriated, reconfigured, and disturbed.

The notion that ancient Hindu scripture held the key to historicizing humans was common to British Indian theories at the outset of the nineteenth century and to Indian theories at its close. But while the Calcutta orientalists read the Vedas as vital evidence for tracing the physical movements and forms of humanity, many of the leading thinkers in the same city around a hundred years later found that these very texts obliterated the value of such inquiries. This case was one of many instances of profound shifts within apparent continuities in nineteenth-century studies of human origins and diversity that focused on the Indian subcontinent. British and Indian thinkers alike drew inspiration from the collisions of texts, materials, and people that marked colonial domination. There was not a smooth exchange of information but, rather, a series of fraught encounters marked by elisions, appropriations, and battles over epistemic authority. The processes of "calibration" that all actors in the colonial Indian subcontinent—not just British and Bengali elites—undertook in order to render information from others commensurable with their existing knowledge structures are certainly worthy of further work in the future.[96]

Reworkings of—and contests over—knowledge took place not between homogenous groups of "colonizers" and "colonized" or "Westerners" and "Indians." They instead involved more specific interactions, often between agents and institutions within, not just across, these categories. Such wranglings attempted to establish the claims of particular people, institutions, places, and belief systems to have privileged insights into the development and variation of humanity. As such, they influenced and were influenced by dynamics particular to colonial India and often partly distinct from the concerns that molded contemporaneous British and European studies on related issues. Nonetheless, many theories of human diversity centered on India—those from Jones to Aurobindo, from Rammohan to Risley—had substantial impacts well beyond the Indian subcontinent. Knowledge did not simply seep from imperial centers to colonial fringes. Instead, what counted as a center was disputed, and the transmission and recasting of ideas took place throughout fragile but far-flung networks, taking in parts of the Indian subcontinent and extending throughout and beyond the British Empire.

During the nineteenth century, data and theories from India morphed to tell a diverse range of stories of the origins and variations of humankind that did not resolve into a singular, authoritative account. Inhabitants, and a host of outsiders, unearthed evidence and received prompts for linguistic and physical methodologies, for monogenist and polygenist contentions, and for bewilderingly long and problematically short human chronologies. The circularities, doublings-back, and overlaps between these competing ideas are hinted at in Herbert Risley's claim that "man alone passes and leaves nothing," at the opening of this chapter. Risley's resonant fragment within his masterwork of physical anthropology could easily have been a spiritualist claim from the likes of Vivekananda on the ephemerality of humanity's physical form and its irrelevance within a grandiose cosmic scheme. In both cases—as in many others in India throughout the nineteenth century—we glimpse human diversity theories at their contested limits.

6

HOW AND WHY DARWIN GOT EMOTIONAL ABOUT RACE

GREGORY RADICK

In *The Expression of the Emotions in Man and Animals*, published in 1872, Charles Darwin claimed to show that, around the world, humans of every race express their emotions in the same ways: crying when sad, smiling when happy, and so on. For Darwin, this sameness afforded what he called a "new argument" for the common descent of all the human races from a single ancestral stock. What follows is a new account of the origins of the empirical research underpinning this argument, including the bravura deep-time historical reconstruction with which Darwin fleshed it out. Understanding how and why Darwin first began to collect evidence on emotional expression across the human races throws new light on the recently controversial questions of when and where, if anywhere, his scientific work reflected his lifelong hatred of black slavery. It also suggests a new solution to an old puzzle: the notoriously "non-Darwinian" character of Darwin's explanations in the *Expression*.

As I will be emphasizing what is distinctive about the *Expression* within Darwin's oeuvre, it is well to notice at the outset a resemblance between one of the most startling statements in that book and one that Darwin makes in the

Descent of Man (1871), not about emotional expression but about the moral sense. In the *Descent of Man*, we read:

> I do not wish to maintain that any strictly social animal, if its intellectual faculties were to become as active and as highly developed as in man, would acquire exactly the same moral sense as ours. In the same manner as various animals have some sense of beauty, though they admire widely different objects, so they might have a sense of right and wrong, though led by it to follow widely different lines of conduct. If, for instance, to take an extreme case, men were reared under precisely the same conditions as hive-bees, there can hardly be a doubt that our unmarried females would, like the worker-bees, think it a sacred duty to kill their brothers, and mothers would strive to kill their fertile daughters; and no one would think of interfering.[1]

Compare the following from the *Expression*:

> If the structure of our organs of respiration and circulation had differed in only a slight degree from the state in which they now exist, most of our expressions would have been wonderfully different. A very slight change in the course of the arteries and veins which run to the head, would probably have prevented the blood from accumulating in our eyeballs during violent expiration; for this occurs in extremely few quadrupeds. In this case we should not have displayed some of our most characteristic expressions. If man had breathed water by the aid of external branchiae (though the idea is hardly conceivable), instead of air through his mouth and nostrils, his features would not have expressed his feelings much more efficiently than now do his hands or limbs.[2]

If man were raised as bees are, if man had breathed water as fish do . . . How striking that, when glossing his reconstructed deep-time histories for humankind, and so guiding the reader on how and how not to interpret them, Darwin availed himself of imagery that was not merely colorful but colorfully counterfactual. Just about everything, he insists, could have turned out rather differently. I shall return to this point—and to its bearing on this volume's concern with the historicizing turn in nineteenth-century thought and its imperial contexts—at the end.

CONSIDERING THE *EXPRESSION*

A "New Argument" for the Unity of the Human Races

The passage just quoted comes from the penultimate paragraph of the *Expression*. Before that we have, in order, an introduction, three chapters setting out the three "principles of expression" that Darwin will use to explain why particular movements came to be expressive of particular emotions, two chapters doing just that for emotional expressions in nonhuman animals, eight chapters doing the same for human emotional expressions, and the bulk of the concluding chapter. When Darwin discusses particular emotions and their expression in humans, he typically takes on three main tasks, in varying orders. He describes the psychology of the emotional state, together with the anatomy and physiology of the movements expressing it (sometimes with reference to evidence from, among other sources, babies and the insane). He explains how, on some combination or other of his principles, those movements became linked to the emotional state that they now express. And he surveys responses to his globally circulated *Queries about Expression*, to check how uniformly, or otherwise, the human races express that emotional state in that way.[3]

To illustrate: in Darwin's pages on anger and indignation, we learn that the moderately angry or indignant man—as distinguished from the enraged man—will experience an increased heartbeat, show heightened color and bright eyes, flare his nostrils, compress his mouth, frown, hold his head erect, expand his chest, plant his feet firmly on the ground, and either square his elbows or hold his arms rigidly by his side. By this point in the book (chapter 10) Darwin has said so much about the role of habit in generating links between emotional states and their expressive movements—"old habits die hard" is a fair summary of his first expression principle—that he evokes it here in explanation with minimal fuss. All a man needs to do today is to imagine he has been insulted, and he will (pointlessly) assume the complex posture of indignation, because he is descended from men who, when feeling indignant, prepared to attack the offender and who, on attacking, felt relief from indignation. If that man is from Europe, where they fight with their fists, then an indignant state of mind will likely lead to clenched fists as well. Everything minus clenched fists, Darwin reports, is how other men, from other races and places, express indignation, including Australians, the Malays, the Abyssinians, the indigenous people of South Africa, Dakota Indians of North America, the Fuegians (about whom more later), the Maori, the Chinese, and low-caste Bengalees.[4]

So it goes, through expression after expression. Only in the book's final chapter, "Concluding Remarks and Summary," does Darwin reveal that the robust uniformity found across the races adds up, in his view, to something important: "a new argument" for belief in the common evolutionary ancestry of the human races. "I have endeavoured," wrote Darwin, "to show in considerable detail that all the chief expressions exhibited by man are the same throughout the world. This fact is interesting, as it affords a new argument in favor of the several races being descended from a single parent-stock, which must have been almost completely human in structure, and to a large extent in mind, before the period at which the races diverged from each other."[5] We will shortly look more closely at the reasoning that, for Darwin, links what he considers the now established fact of cross-racial sameness of expression (the premise of his argument) and the inference to the races' shared ancestry from a "parent-stock" very much like present-day humans in body and in mind (the argument's conclusion).

A brief comment is in order first, however, about the argument's newness. In the *Descent of Man*, Darwin had already argued not only for the common ancestry of the human races from a near-human progenitor but had even indicated how he saw the forthcoming work on emotional expressions as figuring in. Anyone persuaded on general grounds of the truth of the evolutionary theory, he explained, could not but look at the various human races and conclude that they descend from a single ancestor, for the races show close similarities in many and diverse characters, and shared inheritance from a common ancestor is far more probable as the explanation of such a pattern than that each race somehow acquired all of those similar characters separately. In the case of emotional expression, it was not, he contended, mere closeness that the races showed but sameness, identity.[6]

Darwin spells out this same reasoning in the *Expression*, but with a twist: he also considers, and rejects, the theory of natural selection as a possible alternative explanation. On the whole, natural selection is conspicuous by its absence from Darwin's explanations in the *Expression*, so much so that a small scholarly subliterature has emerged with the aim of making sense of what is, in this respect, a strikingly "non-Darwinian" element of the book.[7] Not every reader of Darwin, it must be said, feels the force of a mystery here worth addressing. After all, Darwin was explicit in the *Descent of Man* about his regret at having overdone it previously in ascribing, à la the design theologian William Paley, adaptive value to the characters of organisms, and so having overdone it in attributing the origins of those characters to Darwin's favored explanation for design without a Designer, natural selection.[8] Furthermore, he always accepted roles for other evolutionary agencies, including the one that bulks largest in the *Expression*, the inherited effects of habitual action (so-called Lamarckian inheritance).[9] Even so, the question of what Alan Fridlund called the *Expression*'s "anti-Darwinism" is not a question *malposé*.[10] For one thing, Darwin's attitude toward the theory of natural selection and its explanatory reach in the *Descent of Man* was, his apologia notwithstanding, far from chastened or wary. He went to some lengths to document the evidence in the present for humans occasionally showing inherited variation and struggling under Malthusian conditions, the better to make his case for his reconstruction of the emergence of "man" as a process driven in the first instance by natural selection.[11] For another, Darwinians in our own day find the expression of emotions so obviously well adapted to the demands of survival and reproduction, and thus so easily explained by natural selection, that Darwin's coming to characterize emotional expression as nonadaptive looks the more surprising, even bizarre.[12] Yet that characterization is fundamental to Darwin's project in the *Expression*. The movements that express emotions do not assist in the struggle for life and mates, and they never did. They are not adaptations. They are nonadaptive locked-in legacies from ancestors with bodies liable to form inheritable habits (the first principle), to generate mirror-image versions of those habits (the second principle), and to move in all sorts of odd ways as surges of nervous energy find channels along which to dissipate (the third principle). Natural selection has no purchase.[13]

In the *Expression*, the longest discussion of natural selection emphasizes its irrelevance. The same passage, by no means coincidentally, also supplies the reasoning by which Darwin builds from the premise of cross-racial similarity-unto-sameness in human emotional expression to the conclusion for common ancestry:

> No doubt similar structures, adapted for the same purpose, have often been independently acquired through variation and natural selection by distinct species; but this view will not explain close similarity between distinct species in a multitude of unimportant details. Now if we bear in mind the numerous points of structure having no relation to expression, in which all the races of man closely agree, and then add to them the numerous points, some of the highest importance and many of the most trifling value, on which the movements of expression directly or indirectly depend, it seems to me improbable in the highest degree that so much similarity, or rather identity of structure, could have been acquired by independent means. Yet this must have been the case if the races of man are descended from several aboriginally distinct species. It is far more probable that the many points of close similarity in the various races are due to inheritance from a single parent-form, which had already assumed a human character.[14]

So, to suppose that such numerous and diverse expression-supporting structures were acquired independently, through a process of variation and natural selection in human races descending from "several aboriginally distinct species," is to leave the structural identity unexplained, or at best explained by something "improbable in the highest degree." Common descent from a single already-almost-human "parent-form" is "far more probable."

We should note an important subtlety in Darwin's reasoning here, and more generally in Darwinian reasoning as such. A biological species viewed from a Darwinian perspective is indeed the product of deep-time evolutionary history, through and through. But for purposes of reconstructing that deep-time history, species' characters that are adaptive are problematic. To see the reason, imagine that you are trying to decide whether two species with a particular adaptive character in common are descended from a common ancestor. The character may be a shared inheritance from a common ancestor for which it was also adaptive. Alternatively, the character may have arisen within two separate lineages that, via variation and natural selection, responded similarly—"converged" is the Darwinian term—to similar environments. Absent further information, you have no way to choose between these possibilities, and so no way to reconstruct the evolutionary past with any confidence.[15]

In the *Expression* passage above, Darwin aims to dispel any lingering doubt about whether, with emotional expression, we are dealing with nonadaptive characters. According to Darwin, structures adapted for the same purpose and acquired independently in separate lineages, as in cases of natural-selection-driven convergence, will end up being similar, but not the same. When it comes to emotional expression across the races, however, we are dealing not merely with similarity but with sameness, identity: the signature not of the convergent evolution of adaptive characters but of the common descent of nonadaptive characters. And with the possibility of the reconstruction of the deep-time history of human emotional expression thus secured, Darwin next proceeds to provide it.

An "Idle Speculation" on the Deep-Time History of Human Emotional Expression

"It is a curious, though perhaps an idle speculation," he continues, "how early in the long line of our progenitors the various expressive movements, now exhibited by man, were successively acquired." The note of self-deprecation notwithstanding, what follows is a master class not only in how to reason deep-time historically about humans and other organisms, but how to end a complex book well. For what Darwin now reveals is that, reviewed with deep-time historical questions in mind, the conclusions reached in the previous chapters—themselves ordered, Darwin elsewhere explains, for ease of exposition—can be reassembled so as to yield up the evolutionary history of emotional expression in the human lineage.[16]

His point of departure is the question of whether an emotional expression can be found among living apes and monkeys. If nonhuman primates do it too, Darwin reckons, it must go a long way back, since the most recent common ancestor of apes, monkeys, and humans lived a long way back. And if nonhuman primates do not do it too, then it must be a more recent innovation, begun after the lineage that became the human lineage branched off. Accordingly, Darwin distinguishes between two categories of progenitors: those living "long before they deserved to be called human," in "an extremely remote" or "very early period," and those who were "eminently human," coming "late in the line of our descent." To the former, earlier category Darwin assigns laughing when feeling pleasure; trembling etc. when feeling fearful; screaming etc. when suffering greatly; the making of threatening gestures, accompanied by reddened skin, glaring eyes, and exposed teeth, when feeling rage; protrusion of the lips when feeling sulky or disappointed; arched eyebrows and opened eyes when feeling astonished; and retching movements around the mouth when feeling disgust. To the latter, later category, he assigns weeping when suffering; raised eyebrows and downturned mouth when feeling grief and anxiety; frowning; the posture of indignation and its reverse, the posture of impotence (Darwin's second principle of expression, antithesis, can be paraphrased as "reverse habits come for free"); widely opened mouth etc. when feeling astonished; the lowering of eyes and turning away of the head when feeling contempt or disdain; and blushing when feeling self-conscious.[17]

If the speculation is idle, there is nothing lazy about the reasoning behind it. Consider Darwin's discussion of where to insert weeping into the evolutionary-expressive grand chronology. As set out earlier in the book, he takes weeping to have become expressive of human suffering thanks to the interlocking workings of the human respiratory and circulatory systems. When an ape baby suffers, it screams, thereby bringing relief in the form of grown-up aid and so becoming habitual—Darwin's first principle. The same thing happens when a human baby suffers. But the human baby's scream has physiological knock-on effects with no counterpart in the ape baby. The violent outrush of air from the human baby's mouth results in the blood vessels near the baby's eye becoming engorged with blood. (Darwin regarded this seemingly trivial quirk of human physiology, first discovered by his expression nemesis Sir Charles Bell [1774–1842], as so important that he commissioned further experiments to confirm it from a distinguished Dutch physiologist, F. C. Donders [1818–1889]). On Darwin's theory, this sudden blood bulge near the eye in turn triggers protective contractions of the muscles near the eye, and these contractions in turn put pressure on the lachrymal glands, from which the tears flow. So, weeping awaited not merely the branching off of the human lineage, but the emergence within that lineage of the near-the-eye blood vessels, the muscles around the eye, and a causal link between abrupt expiration and the surging of blood through the vessels—an undoubtedly late development, Darwin concludes. (Unless, he allows, as an improbable but not impossible scenario, man shares a most recent common ancestor with certain monkeys that, though not closely related to man, do weep—in which case weeping will turn out to be as old as anything in the human expressive repertoire.)[18]

The shift of our ancestors from the trees to the ground and the specialization once there, under the influence of natural selection, of feet for upright walking and hands for dexterous manipulating are major features of Darwin's reconstructed evolutionary history for humans in the *Descent of Man*; it is the more fascinating to watch him, in the *Expression*, reason through the consequences for human emotional expression. Some inferences were straightforward enough: "Our early progenitors, when indignant or moderately angry, would not have held their heads erect, opened their chests, squared their shoulders, and clenched their fists, until they had acquired the ordinary carriage and upright attitude of man, and had learnt to fight with their fists or clubs. Until this period had arrived the antithetical gesture of shrugging the shoulders, as a sign of impotence or of patience, would not have been developed. From the same reason astonishment would not then have been expressed by raising the arms with open hands and extended fingers." Others, however, were less straightforward. In Darwin's view, frowning probably awaited the arrival not just of the face-muscle infrastructure behind weeping but of an upright posture as well, in line with his singular connecting of the explanatory dots between frowning and sun glare as a bother to upright man: "[T]he habit of frowning seems to have been acquired chiefly from the corrugators being the first muscles to contract round the eyes, whenever during infancy pain, anger, or distress is felt, and there consequently is a near approach to screaming; and partly from a frown serving as a shade in difficult and intent vision. It seems probable that this shading action would not have become habitual until man had assumed a completely upright position, for monkeys do not frown when exposed to a glaring light."[19]

The book's penultimate chapter concerns blushing, and blushing is also where Darwin's evolutionary reconstruction ends. How did shame, shyness, and modesty come to be expressed by the reddening of parts of the upper body, principally the face? In the blushing chapter, Darwin pours scorn on those who proposed that blushing is the Designer's means for ensuring moral conduct in His most favored species. Among other defects, according to Darwin, that hypothesis fails utterly to account for a number of the facts about blushing, such as its occurring even in the dark-skinned races (where it can be observed only in scar tissue or in the rare albino individual), and often in morally blameless young people who suffer from intense shyness—their suffering compounded, of course, by the unwelcome extra attention that their blushing attracts. For Darwin, the link with self-consciousness is the clue to a better explanation, along with a curious bit of human physiology whereby self-attention to a part of one's body can bring about the relaxation of the muscles around the arteries there. We blush when and as we do, he suggests, because of the combination of the inherited effects of habit and association (his first principle), the tendency of an emotion-generated rush of blood to flow through accustomed channels (a variant on his third principle, roughly "extreme emotion produces excess motion"), and the acute sensitivity that humans have to being judged negatively by others, on our looks and on our conduct. The young are typically more sensitive to such judgments than the old, and women more sensitive than men; blushing is accordingly greater or lesser. But whatever the individual differences, such sensitivity is a species character, and a distinctive one—an upshot of the uniquely human moral sense, itself the upshot of the evolutionarily novel (and relatively recent) conjunction in one primate lineage of high intelligence with high sociality. "Therefore we may conclude," wrote Darwin, bringing his deep-time historical reconstruction to an end, "that blushing originated at a very late period in the long line of our descent."[20]

There follows a sort of coda on how and how not to think about this deep-time history, from which I quoted earlier. Do not suppose, Darwin warns, that the human repertoire of weeping, blushing, and the rest is where emotional expression was heading all along. That "long line of our descent" has not been a kind of biological train track, fitted with rails to keep movement firmly progressing in the direction of humankind as we know it. Some of the most characteristically human emotional expressions happen at all only thanks to what are incidental, even accidental, details of the makeup of organ systems that function mainly to keep us alive. Those details could have been different with no consequences for that functioning, yet with huge consequences for emotional expression. Thus, Darwin points out that most other quadrupeds manage just fine with a different arrangement of blood vessels in the head than humans have; but without our particular arrangement, there would be no weeping. He also, from the other direction, draws attention to an expressive capacity we might well have had but do not: moveable ears. Somewhere along our line of descent, these got lost, along with, Darwin reckons, other capacities, including the ability to vomit at will. Deep-time history, in Darwin's hands, is deeply contingent history.[21]

RECONSTRUCTING DARWIN'S RACIAL-RECONSTRUCTIVE INQUIRY INTO EMOTIONAL EXPRESSION

The Notebook Period, Late 1830s–Early 1840s

When did Darwin first consider collecting evidence on human emotional expression across the races? In the *Expression*, as we have seen, that evidence serves as the empirical foundation for Darwin's "new argument" for the human races' sharing a common ancestor possessing something near to the complete human emotional-expressive repertoire; and the conclusion of that reconstructive argument in turn enables a further, vastly escalated reconstructive argument, the "idle [but non-lazy] speculation" on the deep-time evolutionary history of emotional expression in the human lineage. To anyone familiar with the outlines of Darwin's life and work, there are two obvious places to look for the beginnings of the cross-racial evidence project. One is the "notebook period"; that is, the period of private, London-based theorizing between the opening of Darwin's first private transmutation notebook in the summer of 1837 and his summing of that theorizing in the *Sketch* of 1842, on the view that so many of the ideas of Darwin's maturity (including his ideas on emotional expression) can be found in his notebooks that the cross-racial evidence project is a good bet to be found there too. The other is in or around 1867, the year in which Darwin's *Queries about Expression* began to circulate in handwritten and published form.[22]

Let us take the notebooks first. The ones known, after Darwin's labels, as "M" and "N" carry the bulk of his expression theorizing.[23] In the former, opened in mid-July 1838 and filled by early October that same year, we find perhaps the most famous entry from Darwin's notes on "man," scribbled on 16 August: "Origin of man now proved.—Metaphysic [i.e., theory of mind] must flourish.—He who understands baboon would do more towards metaphysics than Locke" (M84e).[24] The previous month's entries cover very heterogeneous terrain. But everything bears, in one way or another, on the many and varied connections between the mental and the bodily in humans—connections whose existence Darwin took to support what he called "materialism" (e.g., at M19), and whose sometimes very strange nature he delighted in revealing as illuminated by comparisons with what he took to be our animal kin. He wrote in the initial entries, for example, about a man whose odd muscle twitches and manner of holding his hands were remarkably like those of a father dead too long before to have been imitated (M1–2); an ill old woman who, out of nowhere, sang a tune forgotten since her childhood, suggesting to Darwin not merely that song memories "can thus lie dormant, during a whole life time" but that such memories can be likened to the instinctive, unconscious song-knowledge of birds (with Darwin adding that the woman's act of remembering is really better described as "an habitual action of thought-secreting organs, brought into play by morbid action" [M7–8]); the graded scale that separates healthy people from the insane, indeed healthy people from their insane selves—for everyone, Darwin's doctor father told him, is insane sometimes (M13); and so on. For Darwin, evidence from cases of insanity, piled up in entry after entry during that first month, pointed to the existence in humans of trains of thought and action independent of the will and of conscious awareness, and so analogous to instincts in animals. Watching baboons did more for the student of the human mind than reading John Locke (1632–1704), because, contra Locke, the human mind is not a tabula rasa, but is chock-full of inborn, action-influencing ideas, transmitted via inheritance down a primate lineage that includes the common progenitor of humans and baboons.

Emotional expression comes up in the M notebook as yet another class of evidence along these lines—evidence that much of what humans think, feel, and do is not under conscious, willed control but is habitual, instinctual and/or inherited, with origins sometimes to be found in the very distant past. On 16 August, there are two entries before the baboon/Locke entry: the first (M83e) on "heredetary mind" in Darwin's own family ("My handwriting same as Grandfather.—"); and the second (M84e) querying Edmund Spenser's description in *The Faerie Queene* of a rageful character as "*pale & trembling.* & not as flushing & with muscles rigid.—How is this?" Darwin's question here, of what movements do and do not characteristically express a particular emotion, marks the beginning of this line of inquiry for him. In his next entry after the baboon/Locke one, he noted: "Seeing a dog & horse & man yawn, makes me feel how much all animals are built on one structure" (M85). Five days later, on 21 August, he set out his first extended reflections on emotional expression. It is no longer a wonder, he wrote, that humans find it so difficult to hide their emotions—to look tranquil after being insulted, or humble when feeling smugly self-satisfied; or not to laugh when feeling amused, or yawn when feeling bored, or scream when in pain. They are so hard to disguise, Darwin reckoned, because they are so ancient (M92–96). He went on to write a great deal about emotional expression, much of it presenting in embryonic form what will appear much more fully developed in the *Expression*. Near the end of the notebook, in the middle of a long series of entries dated 23 September, Darwin spells out the mind-body programmatic meaning he gave to all this initial theorizing on expression: "The whole argument of expression more than any other point of structure takes its value. from its connexion with mind, (to show hiatus in mind not saltus between man & Brutes) no one can doubt this connexion" (M151).

So, for the Darwin of the M notebook at least, emotional expression mattered for what it could do in tying minds to bodies and, therefore, in tying man to the brutes. There is nothing about what expression could do in tying the human races together, as descendants of a common, nearly human ancestor. This issue, it appears, was just not an issue for Darwin. Different races of humans do make their appearances in the M notebook here and there. We read, for example, that "negroes" seem to have one instinctive notion of beauty and "Europeans" another (M32); that conscience seems to be stronger in some races than in others, though a moral sense is probably universal in humans (M76); and that "whether in Ancient Greeks, / with their mystical but sublime views, or the wretched fears & strange superstitions of an Australian savage or one of Tierra del Fuego" one sees signs of a universal God sense—of innate knowledge of the Creator, implanted by Him when He created "man," as a species separate from all others—Darwin rather doubted (M136–37). If we look to the N notebook, filled from early October 1838 to early August 1839, we find much the same sort of thing.[25]

Again, the rule with Darwin's notebook theorizing on expression—as on the range of topics that engaged him—is that the big public ideas will be found somewhere in the private notebooks. We find in the M notebook clear statements of what became Darwin's first two explanatory principles of expression, and maybe even his third, too—an important point in making sense of the "non-Darwinian" character of these principles, since they thus predate Darwin's famous encounter with Malthus's *Essay* in late September 1838 and so predate the theory of natural selection that, over the next months, Darwin would come to formulate. When, in an M notebook entry from August 1838, Darwin wrote: "Expression is an heredetary [*sic*] habitual movement consequent on some action, which the progenitor did when excited or disturbed by the same cause, which / now / excites the expression" (M107), he was still recognizably under the influence of the evolutionary thinker whose handwriting Darwin believed he had inherited, his grandfather Erasmus Darwin (1731–1802).[26] (Darwin's high regard for the *Zoonomia*, where Erasmus had depicted animal expressions as arising from associations that became habitual and eventually hereditary, dated from his years in medical school in Edinburgh. So did his introduction to the topic of emotional expression as full of potential for those wishing, contra Bell, to stress continuities across the human/brute divide.)[27] The useless uncovering of the canine during sneering ("the very essence of an habitual movement is continuing it when useless" [M96]); pouting and sulkiness in an orangutan at the zoo compared with pouting and sulkiness in humans (M129); the antithetical expression of opposite emotions (second principle) (M147); the expressive role of the discharge of excess nervous energy (third principle) (M150): all are present. Not present, however, in the M and N notebooks, or other extant documents from the notebook period, is a concern to use common emotional expression across the human races to vindicate the unity of the human races. For the emergence of that concern explicitly, we need to go forward in time—but not as far as circa 1867.

The Correspondence of January 1860 (and the Strength of Weak Imperial Links)

In a letter to the geologist Charles Lyell on 10 January 1860, Darwin mentioned sending some questions on expression out to the place where, on the *Beagle* voyage, he had experienced human racial difference at its most extreme:

"I have thought only vaguely on man. With respect to the Races. . . . I have one good speculative line, but a man must have entire credence in N. Selection before he will even listen to it.—Psychologically I have done scarcely anything. Unless indeed expression of countenance can be included, & on that subject I have collected a good many facts & speculated: but I do not suppose I shall ever publish; but it is an uncommonly curious subject.—By the way I sent off a lot of questions the day before yesterday to Tierra del Fuego on expression!"[28] Anyone encountering this letter in the 1860 volume of the magnificent *Correspondence of Charles Darwin* is led to suppose that this letter belongs to a sequence of events in 1860 roughly as follows: (1) Darwin wrote up his questionnaire (6 January); (2) He sent it off to an English missionary in Tierra del Fuego named Thomas Bridges (1842–1898) (8 January); (3) That posting to Bridges got a mention in Darwin's letter to Lyell (10 January); (4) Darwin received Bridges's letter with his answers (October or after); (5) nothing much else happened from then until circa 1867. But this is not what happened, as I will now show in some detail, before turning to an analysis of the autumn 1859 Darwin–Lyell correspondence that, I contend, holds the clues we need to explain Darwin's inventiveness on expression and race in early 1860. If what follows seems, for present purposes, something of an interruptive detour, we do well to take it, partly to resolve the minor mystery of the gap between this 1860 activity around Darwin's expression queries and the much better-known activity from circa 1867 (another "Darwin's delay" case), and partly to seize the chance it offers to begin considering the imperial dimension to Darwin's deep-time historical work on emotional expression.

We start with the item at the head of the chronology above and dated 6 January 1860. In fact, there are two documents, currently held together in the same folder (DAR 185:72–73) in the Darwin Papers in Cambridge University Library: a version in Darwin's hand, one page, front and back (item 73); and a "fair copy," in a hired copyist's hand, two pages front and back, signed at the end (in different ink) "Charles Darwin [/] Down Bromley Kent [/] Jan. 6 1860" (item 72). Both documents have, with one exception, the same text: nine questions on emotional expression in, as the first question makes plain, Fuegians and Patagonians; a couple of how-to guidance notes, urging the gathering of information on any aspect of emotional expression in "savages" ("a subject, which has been wholly overlooked") and the taking of notes at the time of observation, not afterward; and questions about, respectively, the Fuegian ideal of feminine beauty, whether Fuegians "take any pains in breeding or matching their dogs," and the coloring of wild pigs and wild cattle on the Falkland Islands. Atop the version in Darwin's hand only is the title "Expression of Savages." Neither document is addressed to anybody.

The *Correspondence* editors, taking their cue from a scholarly gem of an article on the *Queries about Expression* from 1972 by Richard Freeman and Peter Gautrey (Gautrey later became a *Correspondence* editor), published this text as a letter to Thomas Bridges dated 6 January 1860, on the entirely reasonable supposition that, since Bridges sent back responses to the queries, he must have been their addressee.[29] Although there is no date on Bridges's letter-length reply (nor, for that matter, an addressee or a signature), it too was printed in the 1860 volume, among the October 1860 letters, on the view, as the editors explained, that "the date reflects the minimum time required for Bridges to have received and answered CD's letter of 6 January 1860," though they allowed that "Bridges's letter . . . could have been written later in 1860 or in 1861." They noted finally: "The letter was forwarded to CD by Waite Hockin Stirling, the secretary of the Patagonian Mission Society in London"—and Darwin himself had jotted as much on the bottom of the original, now in DAR 85:39 ("Answer received through Mr Stirling, from the catechist to the Fuegian Mission, Mr Bridges" / "Information from Mr Bridges, Catechist to Fuegian Mission, through Mr Stirling").[30]

From the above, it would indeed seem that Darwin wrote to Thomas Bridges in January 1860, and that Bridges used a fellow missionary, Stirling (1839–1923), as a sort of courier to get the answers back to Darwin. One is left to imagine that, say, Bridges finished up his assignment from Darwin and, not wanting the great naturalist to have to wait longer than necessary, asked a colleague passing through to hand-deliver the response, to the British postal system if not to Darwin himself; and conversely that, for anyone curious about the savages, dogs, pigs, and cattle of Tierra del Fuego in early 1860, the obvious "go-to" correspondent was Bridges. But Darwin in 1860 would have had no idea who Bridges was. The "catechist to the Fuegian mission" was then a twenty-year-old nobody, who had only recently taken over the missionary settlement started by his adopted father in the western Falklands. He became Darwin's man on the spot in Tierra del Fuego and Patagonia—and thus a named source not only in the *Expression* but in the *Descent of Man* (on Fuegian notions of beauty) and *The Variation of Animals and Plants under Domestication* (on the animals)—thanks to a network that Darwin accessed through an old friend from his *Beagle* days, the naval officer and hydrographer Bartholomew James Sulivan (1810–1890).[31]

Sulivan's role as intermediary became clear only with letters published in the 1866 and 1867 *Correspondence* volumes. Unlike Bridges, Sulivan in January 1860 was familiar to Darwin (they had kept in touch) and, thanks to many years spent working and living in the Falklands after the *Beagle* voyage, someone widely esteemed for unrivaled knowledge of the region. That reputation as an expert had landed Sulivan a position in the marine department of the Board of Trade, which was where Waite Stirling, of the Patagonian Missionary Society, had come to see him for advice on setting up a mission in Tierra del Fuego.[32] Now, in January 1860, with Darwin asking a favor from his old shipmate, Sulivan contacted Stirling, who duly passed Darwin's queries on to Bridges. Bridges answered them, and sent the answers back to Stirling, who in turn decided to write a composite letter, combining his own observations with those of Bridges. But the results, Stirling judged, were "so incomplete," as Sulivan explained to Darwin in a letter dated 11 January 1867, "that he [Stirling] did not think them worth sending." Sulivan continued: "On searching his desk he found the questions & answers written by Mr. Bridges which I now send you."[33]

Behind this belated desk-searching by Stirling lies a tale seemingly sprung from Darwin's M notebook—a lost entry on the weird ways of the associative, amnesia-prone, unconsciously active human mind. On Christmas Day 1866, Sulivan, by now retired and living in Bournemouth, sent Darwin a long friendly note, recalling the Christmas they had spent together on the island of Chiloe, and including news about, among other things, the recent doings of the South American Mission Society (as Stirling's society had been recently renamed) in connection with Tierra del Fuego. Like the *Beagle* all those years before, the Mission Society's schooner had recently departed England carrying four Fuegians back to their homeland, where they would soon be joined by Stirling, the man who had brought them over. So deeply involved in this work was Sulivan that he wrote now of "our Mission schooner," "our clergyman there" (meaning Stirling), and the many encouraging signs among the indigenes "that some good influence is beginning to work."[34] Darwin wrote back to say how fascinated and pleased he was to discover that Robert Fitzroy's efforts to civilize the Fuegians had not all been in vain. Then:

> Do you know Mr Stirling well enough to ask him to grant me a great favour? Namely to observe during a few months the expression of countenance under different emotions of any Fuegians but especially of those who have not lived much in contact with Europeans, & to take the trouble to write me a letter on the subject. It is an old hobby-horse of mine on which I am very curious, & on which I have vainly sought for information. I will write a few questions on a separate piece of paper, & if you can oblige me you might send it to Mr S. with the request that he wd hereafter write to me by address on the paper.—[35]

Sulivan did as asked, as he wrote back on 11 January 1867. "I went to Southampton to see Mr. Stirling off," reported Sulivan, "and on giving him your paper he reminded me that I gave him a somewhat similar one from you before—and from his and our catechists [*sic*] notes he had written some answers for you"—the unsatisfactory composite since lost to history, though Bridges's original response was found.[36] Darwin replied a few days later, registering his gratitude to Sulivan, the interest of Bridges's answers to Darwin's questions, and his having "quite forgotten that I had previously sent nearly the same questions."[37]

The delayed arrival of Bridges's response in January 1867 seems to have reenergized Darwin's inquiry into cross-racial emotional expression. He had just sent the *Variation* to the printers, so the project on humans was now coming to the fore, and undoubtedly he would have gotten around to posing again queries that, at some level, he had plainly never quite forgotten.[38] Be that as it may, soon after receiving Sulivan's letter, Darwin drew up a new version of the old questionnaire and began sending out copies: to Brazil (22 February); to New Zealand, China, and South Africa (27 February); and to Australia and the United States (28 February). It was this version that, whether in handwritten, printed, or published form, then made its way around the world, often via chains of correspondence that held fast thanks to the same sorts of personal connections that Darwin exploited through Sulivan, and which the British Empire generated in such abundance.[39] Thus did the acting

Rajah of Sarawak—an Englishman named Charles Johnson Brooke (1829–1917)—become, via Alfred Russel Wallace (1823–1913), Darwin's eyes and ears on expression there in Borneo; an amateur naturalist with a Foreign Office posting in the Far East, Robert Swinhoe (1836–1877), with whom from the mid-1850s Darwin exchanged the gifts that naturalists bestowed (specimens, publications, etc.), become his expert on Chinese emotional expression; and so on.[40] To read in the introduction to the *Expression* Darwin's three-plus pages of thanks to his globally distributed correspondents is to appreciate afresh the strength of the weak ties that made Victoria's Empire, and the "information order" it made possible, so much larger even than the red parts of the map.[41]

How the Autumn 1859 Correspondence with Lyell Concentrated Darwin's Mind on the Need for a "New Argument" on the Unity of the Human Races

So what prompted Darwin to compose his queries in the first place, back in January 1860? His introduction to the *Expression* is quite misleading, suggesting that what really concerned him was whether human emotional expression is innate or conventional:

> It seemed to me highly important to ascertain whether the same expressions and gestures prevail, as has often been asserted without much evidence, with all the races of mankind, especially with those who have associated but little with Europeans. Whenever the same movements of the features or body express the same emotions in several distinct races of man, we may infer with much probability, that such expressions are true ones,—that is, are innate or instinctive. Conventional expressions or gestures, acquired by the individual during early life, would probably have differed in the different races, in the same manner as do their languages. Accordingly I circulated, early in the year 1867, the following printed queries.[42]

On this testimonial, when Darwin started collecting evidence, he had in his sights not the deep-time historical question of the single or multiple origin of those races but the thoroughly ahistorical question of the innate or conventional nature of emotional expression in humans. Taking him at his word, one expects therefore to find him, in the months before January 1860, vexed by doubts about the veracity of the facts that, in his notebook theorizing on the human mind, he had taken for granted as showing expression's innateness, such as the difficulty that people have in disguising what they feel or the appearance in babies of emotional expressions that they are far too young and inexperienced to have acquired on their own (e.g., "Seeing a Baby . . . smile & frown, who can doubt these are instinctive," M96).

Darwin's surviving correspondence for late 1859, however, reveals another story. These are, famously, the months when Darwin was preparing for—and from 24 November dealing with—the publication of the *Origin of Species*. Less famously, they were months largely spent not at Down House in Kent but in the Yorkshire village of Ilkley, the northern outpost of the "water cure" of which the perpetually ill Darwin was so fond. Darwin arrived in Ilkley in early October, shortly after sending in his corrections on the final batch of *Origin of Species* proofs. He reckoned that, after the unremitting labors on the book, his body and mind were ready for a break; and nothing brought him relief like the ice-cold watery treatments on offer at establishments such as Ilkley's Wells House Hydropathic Hotel.[43] But whatever the physical relief from the symptoms that troubled him throughout his post-*Beagle* adult life—and signs are that the Wells House regimen suited him mightily—there was no mental letup. For no sooner had Darwin settled in than he began to receive letters about the *Origin of Species* from its first reader, Charles Lyell. Throughout the summer, as Darwin had finished his proof corrections, Lyell—whom Darwin called his "Lord High Chancellor in Natural Science," since his verdict on the book would count most—had been reading along. Now Lyell delivered his verdict, in a series of extraordinary letters mixing praise and encouragement with, where the stakes were highest, criticism and counterargument.[44]

The Darwin–Lyell Ilkley correspondence is a rich tapestry, deserving of close study as a whole. But for present purposes we can attend to a single thread, concerning, surprisingly enough, dogs. In a handful of brief and fleeting passages in the *Origin of Species*, Darwin suggests that domesticated dog breeds probably derive from several wild canine ancestors. The subject comes up initially in the first chapter, on "Variation under Domestication." His overall aim in the chapter is to impress upon the reader just how variable animals and plants can be under conditions of domestication and relatedly, just how powerful artificial selection, when wielded by the human breeder, can be as a means for producing new varieties out of all of that new variability. Darwin dwells at greatest length on domesticated pigeon varieties and the case for seeing them, in all their amazing diversity, as deriving from a single wild species, the common rock pigeon. Along the way, he pokes fun at those who would declare each domesticated variety to be the descendant of its own, exclusive, aboriginally distinct wild progenitor. But he never goes to the extreme of saying that *all* domesticated races of a certain kind *always* trace back *directly* to a single shared wild ancestor; and he adduces dog breeds as, in his view, a well-attested instance of a group of domesticated varieties with multiple wild origins (though again, he says, do not exaggerate—anyone who thinks anything like a bull-dog can exist in a state of nature needs to think harder).[45] He returns to the subject again in a later chapter, on "Hybridism," noting that, however many different wild canine species were originally domesticated by humans around the world, the separate lineages sprung from those domestications must, over time, have acquired the capacity to interbreed—a point that, for Darwin, underscored the chapter's larger theme of sterility between lineages being present in different degrees in different times and places, in a manner that his theory of gradual species change by natural causes made intelligible.[46]

For Darwin, then, dogs, though not like pigeons in probably tracing back to several wild progenitor species rather than to just one wild progenitor species, were in no way a challenge or a problem for his larger agendas. For Lyell, however, that is exactly what dogs were. Repeatedly, Lyell warned Darwin that his multiple-origins material on dogs threatened to undermine his general argument for a family tree of life, and with it the case for a family tree of humans.[47] In a letter of 22 October, Lyell put it like this: Those who today contemplate the differences between human races as different as "the European, Negro, Hottentot & Australian" will find their perplexity in no way eased on being told that they trace back to several separate parent-stocks. Whereas on being told, in accord with Darwin's "system," that "there was some common ancestor of all these races, as of the greyhound, pug, shepherd's dog &c.," perplexity dissipates. So why spoil matters by endorsing, as Darwin seemed to do in the *Origin of Species*, the hoary old multiple-origins-for-dogs account, associated with the eighteenth-century German-Russian naturalist

Peter Simon Pallas (1741–1811)? In Lyell's view, Darwin really needed to tone down the endorsement, ideally by making it plain that for him all the different dog varieties ultimately descend from a single-but-variable wild canine species, probably a species of wolf. "If this be all," wrote Lyell, "then it shd be distinctly declared that neither the dog nor Man are any more derived from several aboriginal or wild species than other plants or animals having a wide range of races. *Whatever you yield in regard to the dog you will have to concede to every variable species of plant or animal (wild or cultivated) Man included.*"[48]

Darwin wrote back a few days later. Conceding that there was, in Lyell's phrase, "an uncomfortable indefiniteness" in the hypothesis that separately domesticated lineages of dog had gradually lost their sterility when crossed, Darwin nevertheless reaffirmed his stated view that something like this was probably what happened. He went on to say that, in the bigger picture, provided one was convinced (as he and Lyell both were) that jackals, foxes, wolves, and other wild canine species themselves all share a common ancestor, then disagreement about the details of exactly how the domesticated dog breeds came about was neither here nor there. "It is," wrote Darwin, "a curious, but not important subject for us: we"—that is to say, we common-ancestry men, who trace all Xs, however diverse, back to a progenitor X, all Ys back to a Y, and so on—"believe that all canine species have descended from one parent; & the only question is whether the *whole* or only a part of [the] difference between our domestic breeds has arisen since man domesticated them." In other words, taking for granted common ancestry as always the ultimate explanation, can we plausibly account for all the diversity in domesticated dog breeds by supposing that just one wild canine species was bred from by humans (as with pigeons)? Or do we need to suppose that more than one canine species, and so more than one act of domestication by globally spreading humans, was involved at the start of dog domestication? Whichever way the decision goes, we can, in Darwin's view, consistently continue to believe that all dogs, in all their diversity, belong to a single family tree, tracing back to a common ancestor.[49]

The debate continued on and off for the next month. Neither Darwin nor Lyell budged from his original position. If anything there was extension and entrenchment, with, for example, each man treating the comparative evidence on gestational periods in wolves and domesticated dogs as supporting his own position. There was also more emphatic restatement of their views on whether Darwin's endorsing Pallas on dog origins—albeit Pallas "interpreted by the Darwinian key," as Lyell nicely put it in the 22 October letter—created trouble for the larger Darwinian theory by supplying those who denied common ancestry with critical ammunition. "I cannot help thinking," wrote Lyell at one point, "that by taking this concession, one which regards a variable species, about which we know most (little tho' it be), an adversary may erect a battery against several of your principal rules."[50] In the final letter of the exchange, on 23 November, Darwin wrote back, somewhat exasperatedly, that while he would "*infinitely* prefer" the history of domesticated dogs to mirror exactly the simple divergence-from-a-common-ancestor pattern found elsewhere in the tree of life, at the micro- as well as the macro-scale, alas, the facts about dogs suggested a more complicated pattern, and so he had no choice but to acknowledge this local complication.[51]

In sum, throughout October and November 1859, in the run-up to the publication of the *Origin of Species*, Lyell alerted Darwin to the possibility that by allowing for the multiple origins for domesticated dogs, he had weakened his case for common ancestry as a general feature of the history of life, with consequences not least for the history of the human races. "Whatever you yield in regard to the dog," warned Lyell, "you will have to concede to every variable species of plant or animal (wild or cultivated) Man included." We shall come shortly to consider why, in the context of wider developments, this warning was not one that Darwin would have taken lightly. Be that as it may, within the surviving autumn 1859 correspondence, the human races do not figure in a big or even a small way in Darwin's letters to or from anyone else or on any other topic.

What, for Darwin, were the options in responding to Lyell? On the dog side of the dog-breeds/human-races pairing (a pairing that remained with Darwin long afterward; it appears at the beginning of his defense of humans' common ancestry in the *Descent of Man*), there was, from Darwin's perspective, no room for concessionary maneuver.[52] But on the human side, where what was needed was not the bending of the argument but its bolstering, so that it stood solidly on its own evidence base, independent of the complications of the dog case, the situation was very different. And as, in Darwin's final weeks in Ilkley, in late November and early December 1859, he reread the *Origin of Species* to put together a list of amendments for an immediate reprinting of the sold-out book, he would have been forcibly reminded that in constructing a case for common ancestry, the most useful characters are the most useless.[53] In nature interpreted by the Darwinian key, it is, again, the nonadaptive characters—existing because they persist, inertially, from ancestors, not because they assist with the struggle for life and mates—that point the way into the past. From the *Origin of Species*'s penultimate chapter: "Adaptive character[s], although of the utmost importance to the welfare of the being, are almost valueless to the systematist. For animals, belonging to two most distinct lines of descent, may readily become adapted to similar conditions, and thus assume a close external resemblance; but such resemblances will not reveal—will rather tend to conceal their blood-relationship to their proper lines of descent."[54] Not, it must be stressed, that Darwin would have needed any reminding here. His insight into the value of nonadaptive characters as clues to ancestry and, conversely, into ancestry as the best explanation of nonadaptive characters went back as far as anything in his theorizing about transmutation—all the way back, on Jonathan Hodge's analysis, to pre-notebook-period, *Beagle*-vintage reflections on the birds of South America, where Darwin's observations of commonalities across similar species in very different environments provoked him into his first, still tentative dissents from the strongly adaptationist doctrine of the independent creation of species as defended in Lyell's *Principles of Geology* (1830–1833).[55]

So for Darwin—back at Down House from mid-December 1859—the search for new evidence to strengthen the common-ancestry case for the human races would, more or less automatically, have meant the search for evidence of the commonality of useless, nonadaptive characters across the human races. That search in turn, given Darwin's conservative, even archival, tendencies when it came to his own previous theorizing, would have sent him back to his M and N notebooks, now within easy reach, and in any case reviewed by him only a few years before (as dated annotations to this effect testify).[56] And in their pages, as we have seen, Darwin had not only identified human emotional expression as a useless human character but had invested heavily and creatively in a body of theorizing that took that uselessness for granted. Thus did this older theorizing—in many respects, and again as we have seen, so "non-Darwinian"—come to have a new significance for Darwin, and to earn afresh its place within his creative scientific life.[57] In late 1859 and early 1860, Darwin dealt with an increasingly voluminous correspondence on his epoch-making book. Even so, at more or less the first opportunity, he drew up his first list of expression queries, had a fair copy made, arranged for them to be sent to southern South America, and let Lyell know that he should watch that space.

How, and How Not, to Connect the New Inquiry to the British Anti-slavery Movement

We might well ask why, exactly, Darwin was sufficiently disturbed by Lyell's allegation to inaugurate a new inquiry in response. Fortunately a good answer lies to hand, in Adrian Desmond and James Moore's 2009 book *Darwin's Sacred Cause: Race, Slavery and the Quest for Human Origins*. Unfortunately the book purports to defend an implausible thesis—namely, that Darwin came up with his evolutionary theory in order to strike a blow against black slavery—and so that good answer risks getting overlooked.[58] No book is a package deal, however, and Desmond and Moore's, so rich in new ideas and original scholarship, benefits from charitable reading and selective appropriation. On grounds of charity I incline to ignore the bolder, programmatic statements of the thesis and concentrate instead on what the body of the book achieves. Putting those achievements to work in interpreting Darwin's 1859–1860 decision to collect new evidence on emotional expression across the human races—and thus to make possible what would become the *Expression*'s "new argument" for the unity of the races, as well as the amazing deep-time historical reconstruction erected on its basis—is a step in the right direction.

To this end, three achievements in particular stand out. First, the book shows how deeply invested Darwin was from early days in the notion that the human races share a common ancestry. One can read a lot of Darwin without picking up on the extent to which he was far from neutral on this question. In the *Descent of Man*, for example, he represents himself as having written a book about humans at all only because he reckoned it was time to apply his general theory in detail to a particular species, and humans seemed as suitable as any. When he comes later in the book to the human races, he accordingly follows a pattern of argument familiar from the *Origin of Species*, bringing on the common ancestry of the races only after first showing that some evidence supports the ranking of the races as mere varieties while other evidence supports their ranking as distinct, different species. We thus meet common ancestry in the human case as functioning no differently than it does elsewhere in the Darwinian oeuvre, as the best explanation for otherwise unresolvable ambiguity in varieties-or-species debates.[59] And again, in the *Expression*, Darwin springs the common-ancestry upshot of his racial evidence on the reader only at the very end, as a kind of surprise bonus, an interesting afterthought.[60]

Behind such artfully displayed dispassion, however, lies a quite different biographical reality, as Desmond and Moore reveal. The common descent of the races—their unity in blood—was an article of faith in the culture of British antislavery in which Darwin grew up, thanks to his family's deep and long-standing involvement. In that culture, moreover, the denial of the unity of the races was associated with the slavers, who clung to it as a rationale for treating black men and women abominably.[61] Darwin's absorption of that set of linkages shines through in a letter that he sent to his cousin W. D. Fox (1805–1880) in 1850. Commenting on a lecture recently given in Charleston, South Carolina, by the Swiss naturalist Louis Agassiz (1807–1873), defending the multiple ancestry of the human races, Darwin wrote: "Agassiz's Lectures in the U.S. [maintain] the doctrine of several species,—much, I daresay, to the comfort of the slave-holding Southerns." Yes, it is a throwaway line in a private letter. But it is all the more telling for just that reason. When his guard was down and the matter of the unity of the human races came up, Darwin showed himself to be every inch a child of the British antislavery movement—a movement whose cognitive as well as emotional legacies Desmond and Moore made vivid as no one before them.[62]

A second achievement of their book is the placing of Darwin within another, related transatlantic debate, over the origins of domesticated animal varieties as bearing on the question of human racial unity. So virtuosic is Darwin's argument in the *Origin of Species* for the common ancestry of domesticated pigeons—we have seen the impression that it made on its first reader, Lyell—that we easily overlook the fact that Darwin saw himself not as putting forth a boldly controversial view but as siding with consensus ("I am fully convinced that the common opinion of naturalists is correct, namely, that all have descended from the rock-pigeon").[63] Strikingly, one of the books that helped to turn that view into common naturalist opinion was titled *The Doctrine of the Unity of the Human Race Examined on the Principles of Science*, and published in, yes, Charleston in 1850, by an American clergyman naturalist named John Bachman (1790–1874). A friend of Darwin's from time spent together in London in the late 1830s, and of Lyell's after Lyell visited on a trip to the States in the mid-1840s, Bachman made explicit the links he saw between his topic and the defense of slavery. Of course, he wrote, the Christian, so eager for the vindication of the scriptural teaching of the unity of the human races, must not let that eagerness overcome his willingness to confront the facts as exhibited by scientific men. But others, with interests in the opposite direction, must take equal care: "The advocates of a plurality of races should especially be on their guard lest the enemies of our domestic institutions should have room to accuse them of prejudice and selfishness, in desiring to degrade their servants below the level of those creatures of God to whom a revelation has been given, and for whose salvation a Savior died, as an excuse for retaining them in servitude."[64] Bachman was no abolitionist; indeed, his wife held slaves. But his unity-defending book is all the more representative of a moment when a range of people who disagreed about a great many things nevertheless regarded as connected (1) debates about the common or plural ancestry of the human races, (2) debates about the common or plural ancestry of domesticated animal varieties, and (3) debates about black slavery and what to do about it.[65] No wonder, then, that in the autumn of 1859 Lyell found it so natural to ask Darwin whether passages in the *Origin of Species* seemingly backing a plural origin for dogs might leave room for his enemies to cast doubt on his larger case for common ancestry—and to associate that larger case in turn with the human races. And no wonder, too, that Darwin, for all that he found Lyell's persistence exasperating, took him seriously.

There is not even an index entry for "slavery" in the 1859 volume of the *Correspondence.* Yet by Desmond and Moore's version of events, slavery was, subtly but pervasively, everywhere. Showing us just how profoundly *bothered* by slavery Darwin was—and how alert he was to its fortunes in the world, and how alarmed he would have been even to be suspected of aiding it—is the third achievement I want to try and bring into focus. We learn from Desmond and Moore, for example, that while writing the *Origin of Species*, as the news out of America made war over slavery seem ever more inevitable, Darwin turned for downtime reading to a book reporting on life in the American slave states. His son later recalled Darwin saying how the horrors he had encountered in its pages had kept him up at nights (though that did not keep him, once the war had started, from recommending it).[66] Or consider the complex way in which, on Desmond and Moore's recounting, slavery and antislavery hovered over the Darwin–Lyell correspondence that autumn. Antislavery had been a Whig cause par excellence, and Lyell and Darwin each were the very picture of the Whig man of science. But Lyell on his trips to the States in the 1840s had become rather fond of the slaveholders he had met and had written sympathetically about their situation. Darwin had rebuked him for it, in a letter to Lyell but also, Desmond and Moore suggest persuasively, in print, in a scorching but naming-no-names passage added to the 1845 edition of Darwin's *Journal of Researches*, a copy of which Darwin sent to Lyell (the new edition's dedicatee). "Those who look tenderly at the slave-owner," the passage runs, "and with a cold heart at the slave, never seem to put themselves into the position of the latter." At the end: "It makes one's blood boil, yet heart tremble, to think that we Englishmen and our American descendants, with their boastful cry of liberty, have been and are so guilty."[67] Accusatory concern about letting the side down on slavery, then, was part of the dynamic between Darwin and Lyell well before 1859. That autumn, briefly, Lyell reclaimed the moral high ground in his implacable harping on the *Origin of Species*'s minor remarks on dogs as giving succor to the enemy. His provocation set in motion the thoughts and feelings that led Darwin, in compensatory mode, to develop a new line of investigation into emotional expression across the human races.

Standing back from the details above, we can, thanks to Desmond and Moore, identify three main ways in which Darwin's belonging to the world of British antislavery helps make sense of his response to Lyell's criticisms in 1859–1860: in instilling in Darwin a deep commitment, as much intellectual as moral, to the view that the human races share a common ancestry; in familiarizing him with the tradition whereby debates over domesticated animals served a surrogate role for origins debates over the human races (and thus over the legitimacy of black slavery); and in sensitizing him to the ongoing evils of black slavery as an outrage needing to be confronted and combated. We miss out on a great deal if we fail to take these contexts—distinct but overlapping—into account in understanding why Darwin acted as he did when he did. And again, to acknowledge the explanatory power of these contexts is not therefore to accept that, in Desmond and Moore's words from their preface, "the British anti-slavery movement . . . is the key to explain why such a gentleman of wealth and standing should risk all to develop his bestial 'monkey-man' image of our ancestry in the first place."[68] I doubt that there is just one key to Darwin, or that Darwin saw his theorizing about transmutation as especially risky for someone of his class (rather, the reverse), or that his antislavery background mattered more to his initial theorizing on species than did his reading of Lyell and firsthand observations on bird biogeography in South America. I would not even say that Darwin's antislavery background is "the key" to understanding the origins of his cross-racial inquiry into human emotional expression. But I think it helps, as nothing else does, in understanding why Lyell's criticisms spurred Darwin to action.

If *Darwin's Sacred Cause* illuminates the *Expression*, the *Expression* returns the favor, as Darwin's racial-reconstructive inquiry can now be seen as bearing the stamp of the British antislavery movement more plainly than does anything in the *Origin of Species* or the *Descent of Man*. Charitably minded readers should, among other indulgences, supply Desmond and Moore's book, which ends in 1871 and the *Descent of Man*, with a new final chapter. In doing so, moreover, they can remedy a further shortcoming by giving due prominence to an uncomfortable passage in the *Expression,* where Darwin claims that his theory well explains why "the children of savages should exhibit a stronger tendency to protrude their lips, when sulky, than the children of civilized Europeans." "Savage children" express a sulky state of mind more strongly, Darwin goes on, because "the essence of savagery seems to consist in the retention of a primordial condition."[69] Even in the *Expression*, site of Darwin's most original contribution to the case for the unity of the human races, he is unembarrassed about his belief in a racial hierarchy. For Darwin, "savages" are closest to our animal-like progenitors, and "savage children" closest of all. Desmond and Moore tend to treat such race-hierarchical moments in Darwin's writings—and there are more than a few—as a sign of Darwin contradicting his better self. But with Darwin, as with that other great antislavery man with whom he shared a birthday, Abraham Lincoln (1809–1865), the challenge is to reinhabit a conception of racial unity that, unlike our own, took racial hierarchy for granted.[70]

CONCLUDING REMARKS AND SUMMARY

I take as my final heading the title that Darwin gave to the final chapter of the *Expression.* As we have seen, Darwin there discharged the traditional end-of-the-book obligation of reviewing the preceding pages by reorganizing his major findings into a deep-time history of emotional expression. Without any illusions about how my efforts compare with Darwin's, I want also now to summarize chronologically what I have so far presented, for expository purposes, out of chronological order.

While Darwin was a medical student at Edinburgh in the 1820s, he encountered the topic of emotional expression as bearing in a subversive, materialist vein on questions about how mind relates to body and how humans relate to nonhuman animals. Unsurprisingly, when he began in the late 1830s to work out the subversive, materialist implications of his new species theory for the understanding of the human mind, he seized upon emotional expression in the familiar Edinburgh spirit, as showing how little control our wills have over our bodies and how much the peculiar connections between our feelings and their expression owe to habits formed either by ourselves or—as comparisons with apes and other animals suggested—by our ancestors. There is scant sign in Darwin's notebooks from this period of an interest in whether the different human races express emotions in the same way, much less in whether the collection of evidence for that sameness might be useful in defending the doctrine that all the races share a common ancestry: a central tenet of the British antislavery movement to which Darwin was heir. The idea of gathering evidence on cross-racial emotional expression emerged for Darwin only some twenty years later, between October 1859 and January 1860, in the course of his correspondence on the *Origin of Species* with Lyell. For Lyell, Darwin's backing of multiple origins for domesticated dog varieties in the *Origin of Species* contradicted—and thus weakened—his general argument for common ancestry, including the common ancestry of the human races. Darwin thought that Lyell was making too great a fuss over what was not even an exception to the rule, just a minor and local complication. But even being suspected of giving "comfort to the Southern slavers," as Darwin once accused the multiple-origins Agassiz of doing, was not something the passionately antislavery Darwin would have taken lightly. He knew, as Lyell did, that over the past twenty years, as the campaign for the abolition of black slavery in the United States had gathered momentum, the common-or-multiple origins of the human races had often been argued in tandem with the common-or-multiple origins of domesticated animal varieties. And Darwin also knew that in his recently reviewed notebooks on the human mind, he had treated emotional expression as just the sort of nonadaptive character required for the reconstruction of descent from a common ancestor. Within six weeks of the publication of the *Origin of Species*, Darwin had a new response to Lyell's accusation: a questionnaire about emotional expression in other races. He sent it in the first instance to Tierra del Fuego, via the same imperial missionary network that had initially taken him to that place of maximally-different-from-him humans. For reasons both mundane and bizarre, he ended up not receiving the answer until early in 1867. But from that point he mounted an ever-escalating attack on the problem, eventually receiving, as he wrote in the *Expression*'s introduction, "thirty-six answers from different observers, several of them missionaries or protectors of the aborigines, . . . [and relating] to several of the most distinct and savage races of man." On their basis, and in particular on what he claimed as evidence for the universality of human emotional expression across the world, he raised, in the book's conclusion, a "new argument" for the unity of the human races and reconstructed the shared deep-time history behind their common emotional expressions.

Empire comes up in two equally important but, as it were, opposite forms in the above. There is, on the one side, those myriad, power-entrenching and power-extending imperial forms so well represented in the *Beagle* voyage, from the coastal surveying that was its official rationale to the article that Darwin and his captain, Robert Fitzroy, coauthored in praise of the Tahitian missionaries and their civilizing work.[71] But there is also, on the other side, that remarkable renunciation of one widespread imperial practice, slavery, by the British movement for its abolition. Without this movement, Darwin would never have come to be so deeply committed to defending the common ancestry of the human races, so sensitive to the charge that his pluralism about domesticated dogs might have imperiled the case for racial unity, and so determined to support that case with fresh evidence. Without the empire itself, there would never have existed that network of global observers ready to serve as Darwin's informants on emotional expression, nor would Darwin have enjoyed the level of access to this network that, thanks to the *Beagle* mate turned Fuegian mission master Sulivan, functioned as an immediate point of entry. We can, if we like, label Darwin's cross-racial expression inquiry, and the deep-time history that Darwin reconstructed on its basis, "imperial science," provided our conception of empire is capacious enough to include its negation.[72]

But however we choose to label it, the inquiry looks, as a product of history, rather fragile. The idea for it came to Darwin relatively late and relatively suddenly. There were no obvious precedents for it in the work of others and no reason to think that Darwin would have invented it absent the pressure that came from Lyell as and when it did (though that it came from Lyell was, given their shared background and their own history on slavery/antislavery, no accident). Darwin's answers to his humanly historicizing questions look no less fragile under closer historical inspection. It has, for example, long been noted that in the expression questionnaire published in the *Expression*, Darwin quietly dropped a question from the version that circulated in the late 1860s, about whether, as "a sign to keep silent, . . . a gentle hiss [is] uttered." The editor of the third edition of the book, Paul Ekman, suggested that maybe Darwin left out shushing "because unlike the others it does not deal with expression or gesture."[73] More recently, however, Hong-Jin Liu, on the strength of a systematic examination of Darwin's correspondence on expression from China and other parts of Asia, has proposed an alternative explanation: that Darwin scuppered the question because the information he received indicated that shushing when wanting to silence others was far from universal. We might well look again at how Darwin handles the evidence he did publish, adroitly acknowledging all sorts of diversity in his discussions of particular expressions (such as indignation, expressed, as we saw, mostly but not exclusively by clenched fists), yet still declaring at the end a sameness that is inexplicable unless the races share a common, nearly human ancestor.[74]

Or consider that insistence on the nonadaptive nature of emotional expression: the premise on which Darwin founds not only his explanatory principles of expression but, ultimately, his case for the common descent of the human races. No sooner has Darwin set out that case, including his deep-time historical reconstruction of how and when our lineage acquired its emotional expressions, than he casually admits that, in all sorts of ways, our emotional expressions are useful in the struggle: "The movements of expression in the face and body, whatever their origin may have been, are in themselves of much importance for our welfare. They serve as the first means of communication between the mother and her infant; she smiles approval, and this encourages her child on the right path, or frowns disapproval. We readily perceive sympathy in others by their expression; our sufferings are thus mitigated and our pleasures increased; and mutual good feeling is thus strengthened."[75] Little wonder that Darwinians in our time, relaxed about the unity and indeed biological equality of the human races (on Darwinian grounds), have found it wonderfully easy to explain human emotional expression as the upshot of natural selection. Nor, in Darwin's own time too, did his historicizing efforts in the *Expression* always persuade. The reviewer for the *Times*, for example, noted waspishly that "whenever Mr. Darwin is in a great difficulty he brings in an early progenitor to cut the knot." "[T]he suppositions of the Ptolemaic system," complained the reviewer, "were a modest contrivance compared with this device," which the reviewer further impugned with words that, in the book, Darwin had aimed at the doctrine of the independent creation of species: "By this doctrine, anything and everything can be equally well explained."[76] That was unfair, but not wildly so. Darwin in the *Expression* did not solve the problem of how best to use evidence from the present in order to constrain conjectures about the deep-time historical human past. But he made a remarkably good—and, when slavery came to mind, clenched—fist of it.

His most outrageous conjectures, as I noted at the outset, were to do not with what actually happened in the deep-time history of humankind but with what might have happened—with, as we would say now, the counterfactual past. The contingency of the history of life, its profound dependence on chanciness and above all the chanciness of which species end up in which environments, was there for Darwin from the start, in those first *Beagle*-era questionings of the arch providentialist Lyell.[77] The *Times* reviewer was on to something, then, in linking Darwin's reconstructive reasoning about human emotional expression with the wider challenge of the whole Darwinian project to the idea of divine micromanagement of earthly happenings. Working out exactly what impact those linked innovations had on historical thinking more broadly is a job for future historians of historicization. For now, we can do worse than notice how far the chanciness of encounter at the heart of Darwin's historicizing of species was, as Jonathan Hodge has stressed, part and parcel of the colonial expansion at the heart of *Beagle*-vintage English capitalism.[78]

A final return to the autumn 1859 correspondence between Darwin and Lyell over the *Origin of Species* furnishes an emblematic moment out of this imperial history of historicity. What finally converted a reluctant Lyell to Darwin's side of the argument on species origins was Darwin's observation, in a letter of 11 October 1859, that over and over again, when Europeans brought their animals and plants to the often very different environments of their colonies, the newly introduced species thrived, to the point of going native. This showed, in Darwin's (and eventually Lyell's) view, that species were not all divinely custom-made for the habitats where they originated, fitting tightly in every detail to conditions in their native locales and to nowhere else. Species were the products not of miraculously providential handicraft but of mundanely contingent historical process.[79] Thus did a lesson of empire become one of the first lessons of a new kind of science—and a new kind of history in the bargain.

7

THE COMPARATIVE METHOD IN "SHALLOW TIME"

Walter Scott, Thomas Carlyle, and Francis Galton

HELEN KINGSTONE

> *It is well known that the mystery which overhangs what is distant, either in space or time, frequently prevents us from censuring as unnatural what we perceive to be impossible.*[1]

Like L. P. Hartley's famous opening line—"The past is a foreign country"—Thomas Macaulay's (1800–1859) words here elide time and space, turning one sort of distance into another and suggesting that we can think of exoticism in one category as equivalent to exoticism in the other.[2] He thus goes some way toward recognizing that curious quirk of nineteenth-century thinking, inherited from the conjectural historians of the Enlightenment but exacerbated with expanding colonization and imperialism: the notion that cultures can be mapped against each other as if through the ages of "man." Anne McClintock has described this evocatively as "panoptical time" and "anachronistic space," arguing that in the wake of Darwinism, human and natural times were mapped onto each other, turning the colonial space into what Edward Burnett Tylor (1832–1917) termed a "survival," which McClintock defined as "an anachronistic moment of prehistory."[3] According to that model, an unknown or alien culture from another part of the world can be understood by being measured against those of the past. This is also apparent in Walter Scott (1771–1832), in his introductory epistle to *Ivanhoe* (1820), when he comments that "it was not above sixty or seventy years . . . since the whole north of Scotland was under a state of government nearly as simple and as patriarchal as those of our good allies the Mohawks and Iroquois."[4] Scott evokes an imaginary graph where nations are plotted in relation to the linear path of progress. Geographically and chronologically distant cultures can thus be seen at equivalent points on the scale—and can be treated accordingly.

Macaulay's statement also highlights another important facet of nineteenth-century thinking on human development: that much of it was founded on conjecture, speculation, and theory. Even once humans had been joined into a trajectory from apes via the Paleolithic and Neolithic periods, "missing links" remained, opening the door to racial speculation.[5] The stadial model evoked by Macaulay—one in which history progresses through set and distinct stages—is an essentially conjectural one that requires an act of imagination as well as one of rationality. This lies in continuity with the conjectural historians of the eighteenth century. For Scottish Enlightenment thinkers such as Adam Ferguson (1723–1816), Adam Smith (1723–1790), and John Millar (1735–1801), stadialism was useful for developing a narrative and analysis of human history and human origins. Without any notion of deep time (James Hutton's theories were not published until 1788) and without any access to prehistoric remains, they had to base their debate about human origins on a different set of principles. As Dugald Stewart (1753–1828) put it, he and others had to "supply . . . the place of fact by conjecture" and draw the most "likely" behavior of earlier peoples "from the principles of their nature, and the circumstances of their external situation."[6] As ideas such as "ontogeny recapitulates phylogeny" highlight, analogy, metonymy, and synecdoche were key elements of any theory of human development through the eighteenth century and into the nineteenth. The same models of envisaging change permeated both "literary" and "scientific" texts, becoming in the latter arguably the more far-fetched and requiring more imaginative investment.

These stadial and comparative models were highly flexible. They could be applied across thousands of years, or within the past half century, to construct stories of human origin that were either monogenetic or polygenetic, and which saw different races either as different points on the same continuous scale, as more or less successful rivals in the "race for life," or even as separate and unrelated species. As Ronald Meek highlights, polygenism could be an "unintended by-product" of the Enlightenment tenet that human nature (though universal) was affected by changes in circumstance: under the same conditions, unrelated tribes could have independently developed in parallel.[7] Roxann Wheeler and Colin Kidd both emphasize that polygenetic theories always existed outside Christian orthodoxy—Kidd even refers to it as "heresy"—and were minority views even in their heyday.[8] It was nonetheless taken up zealously by aspiring nineteenth-century imperialists. As Kidd stresses, "the rise of a racialist paradigm" ran in parallel with "the nineteenth-century crisis of faith": in his image, "inner turmoil . . . seethed beneath the white man's swagger."[9]

What light can be shed on this question of deep time—and its attendant questions of human origins—by examining how the conjectural and comparative methods functioned in relation to much more recent periods: shallow time, if you will? As scholars including David Livingstone and Chris Manias have shown, the "comparative method" was widespread in the nineteenth century among ethnologists and anthropologists.[10] Tony Bennett highlights how, in its revived form, it became a way "to account not merely for the transition from rude to civilized societies but also for the history of the earth and of life on earth."[11] I will show here how a stadial model of historical periodization, which saw humankind progressing upward through set stages and making the transition between stages via transformative watersheds, was widely used to supplement and justify the "comparative method" of judging different civilizations. Discussions of human origins and "deep history" were thus bound up with questions about how to interact with different ethnic groups in the rapidly expanding British Empire.[12]

Stadial structures were widespread in the innovative historicist philosophies of the nineteenth century, which despite being announced as secular nonetheless inherited from the Judeo-Christian tradition a persistent sense of teleology. As John Rosenberg has described it, ideas of the divine were in the nineteenth century not "dissipated so much as transmuted, the Word not eradicated but discovered in new places." The most potent of these "new places" was what Rosenberg terms the "temporal scripture" of History.[13] Georg

Wilhelm Friedrich Hegel (1770–1831), Auguste Comte (1798–1857), and Karl Marx (1818–1883), all envisioned humankind progressing in historical stages, whether from thesis and antithesis to synthesis, from theological to metaphysical states, or from rule by aristocracy to that by bourgeoisie and then proletariat.

In contrast, the gradualist narrative of human history that emerged from nineteenth-century geological discoveries was a disconcertingly "noiseless revolution."[14] John Burrow highlights the shock to the system that these discoveries were to early Victorian thinkers. He shows how after 1830, when Charles Lyell (1797–1875) drew on Hutton's earlier idea of "deep time" and expounded his geological theory of sedimentary gradualism, the geological metaphors used by Victorian writers transformed from those of volcanic eruption to ones "almost invariably sedimentary. . . . Henry Maine uses it for law, Bagehot for society as a whole."[15] This had a significant effect on history writing. Burrow argues that the image of history as a sedimentary process makes it one "whose longer-term significance lay far beyond the knowledge of the actors engaged in it, for it could only be perceived retrospectively and therefore necessarily ironically, though perhaps reverentially."[16] A heightened consciousness of the enormous time scale of geological and, by implication, historical impact made it all the more difficult to acquire an external perspective on historical events. Writers of history were forced to recognize that the ultimate impact of the events they analyzed might remain unknown for a very long time. One way to get around this problem, therefore, and to predict what might happen to a civilization in the future was to plot it on a graph and see where it aligned with a period in Britain's past. Ironically, this made predictions of the future course of British history the greatest challenge of all. If Britain was the most advanced nation in current existence, then there was no precedent against which to compare it. The only possible antecedents were lost and collapsed civilizations: the ancient empires of Egypt, Babylon, Greece, Rome, and Constantinople. Macaulay's gloomy prediction about the inevitable future New Zealander visiting a ruined London comes into sharp focus in this context.[17] Even in his chain of progressive Whig reasoning, there was no way on but down.

It might seem, therefore, that the gradualist model of historical change dominated in the nineteenth century, and it was certainly the preferred model for conceptualizing the history of England, with its distinctively organic

Burkean constitution. As Burrow has shown, the utilitarianism and Whiggism respectively of James Mill (1773–1836) and Macaulay both contributed to the eclipse of conjecture.[18] And as Macaulay demonstrates, space-for-time analogies, even in comparative mode, could draw on gradualist as well as stadial imagery. This is inherent in the Lyellean image of "deep time," where time can be read in the vertical strata of sediment, as it is in Darwin's use of different Galapagos islands and their respective species to signify different points in a gradualist evolutionary process.

What, then, is the link between the late eighteenth-century conjectural history and the late nineteenth-century comparative method? Rather than stadial history being eclipsed after the 1790s before a revival in the late nineteenth century, as Bennett claims, this model of history was surprisingly resilient through the period. A stadial model persisted for the purposes of comparative history. This suggests that we need to see the stadialism of Hegel, Comte, and Marx not as isolated oddities but as part of an embedded and continuing wave of stadial theorizing that both preceded and went beyond them and was often used for ends other than those they championed. In the face of geological (and later, Darwinian) arguments for sedimentary gradualism, those who used stadial models persistently dredged up and gave prolonged credence to a catastrophic and dramatic narrative of overall human history. Peter Bowler, following Stephen Jay Gould, has emphasized interplay and competition between "the progressionist and the cyclic" views of development; I would like to draw attention to another related but distinct competition between models: the gradualist versus the stadial and implicitly catastrophist.[19]

Walter Scott uses the stadial-comparative model in a (self-)congratulatory way in his novels of the 1810s, while Thomas Carlyle (1795–1881) in the 1840s and Francis Galton (1822–1911) in the 1860s deploy the same model to deal with and propose solutions to matters of anxiety. Their mutual use of Scotland as a point of comparison is no mere coincidence. It was in Enlightenment Paris and Edinburgh that the conjectural theory of history was first developed and where it was most applied and most needed—because, as Meek highlights, the life of these cities "was being rapidly and visibly transformed . . . as a result of profound changes taking place in economic techniques and basic socio-economic relationships."[20] Scott was steeped in the ideas of the Scottish Enlightenment historians, and indeed among the elements that made Scott's novels so compelling was their underpinning by a stadial model of history and also their dramatization of the struggles undergone at points of transition between stages.[21] Robert Chambers (1802–1871), the Edinburgh publisher and anonymous author of the popular *Vestiges of the Natural History of Creation* (1844), was an acquaintance of Scott's and unsurprisingly steeped in his ideas.[22] Scotland had undergone dramatic changes through the eighteenth century, propelled particularly by the Darien disaster, the consequent 1707 Act of Union, and the clampdown on Highland culture after 1745 in conjunction with a flourishing Enlightenment in the Lowlands. The phenomenal success of the *Poems of Ossian* (1773), even after they were revealed to have been written rather than merely "translated" by James Macpherson (1736–1796), highlights the appetite in late eighteenth-century England for revelations of the sublimely "primitive" nature of Scottish culture.[23] Scotland thus provided a perfect case study for stadialism, as Scott, Carlyle, and Galton exemplify.

My wide-ranging case studies also show how fluid the boundaries were between literature and science up to the late nineteenth century. As the diversity of these three writers demonstrates, the ideas of comparative history and the stadial models used to propagate them were far from being esoteric or obscure. Theodore Koditschek has engaged with these ideas in the work of Maria Edgeworth (1768–1849), J. A. Froude (1818–1894), and Robert Knox (1791–1862) among others; I look more closely at selected writings—similarly in the form of novels, periodical diatribes, and scientific tracts—by writers who variously initiated or responded to the ideas of those writers. In this chapter I also contribute to the continuing debate, driven in particular by Peter Bowler's work, about the extent to which Charles Darwin's theories had been fully assimilated by the end of the nineteenth century. Bowler has argued for the continuing influence of non-Darwinian, primarily teleological, forms of evolutionary thinking in the latter half of the century.[24] The writers examined here generally assume their stadial models of human history to be upwardly progressive and/or teleological. These two terms do not, of course, have to be synonymous. As Macaulay's New Zealander epitomizes, teleology can imply deterioration of one's own society. Those who took the view that "ontogeny recapitulates phylogeny" and who paralleled (as in the medieval conception) the "seven ages of man" with the "seven ages of the world" inherently saw time as deterioration, just as maturity in human beings can all too quickly be rewritten as ageing.[25] Perhaps the best time was in the past, or fast slipping away? It is notable that those who developed the four-stages theory always placed their own society in the fourth and final stage. What we see, in the case studies that follow, is a combination of these teleological anxieties (has our civilization already reached its zenith?) and faith in a Spencerian evolution-as-progress (as other civilizations are still rising through the stages, the best is yet to come). In the work of these three writers, we can see non-Darwinian forms of evolution both predating 1859 and continuing after it.

WALTER SCOTT AND SCOTLAND

We can see how widespread these stadial models were in the early nineteenth century by their repeated use in Walter Scott's novels. Scott popularizes the use of stadial history as a means of comparison between different cultures. He persistently markets his historical romances (even those set at a comparatively small chronological remove) through highlighting the dramatic stadial transformations that Scotland had passed through between the time of the setting and that of writing. In his first novel, *Waverley; or, 'Tis Sixty Years Since* (1814), Scott explains the apparently alien culture of his 1740s Highland setting, by recourse to comparative stadial models. At the close of the novel, he draws a trajectory of almost unprecedented transformation between its time of setting and that of its first readers: "The gradual influx of wealth, and extension of commerce, have since united to render the present people of Scotland a class of beings as different from their grandfathers, as the existing English are from those of Queen Elizabeth's time."[26] Here sixty years are equated with three hundred. The speed of transition from one stage to another might vary, he claims, and Scotland has undergone transformation at a spectacular rate. This additional dimension enables greater flexibility and thus more varied comparisons in the stadial model.[27]

Scott uses this in part as a defense of his novel's wild and romantic plot and to lend legitimacy to his use of eyewitness accounts as sources. His research for the novel involved speaking with people who could remember its key event, the Jacobite Rebellion of 1745, even though he himself had not lived through it. He situates his novel on the edge of but just about within living memory. In the preface to his first medieval novel (and first set in England) *Ivanhoe* (1819), he asserts that "all those minute circumstances belonging to private life and domestic character, all that gives verisimilitude to a narrative, and individuality to the persons introduced, is still known and remembered in Scotland; whereas in England, civilization has been so long complete, that our ideas of our ancestors are only to be gleaned from musty records and chronicles."[28] As well as a defense, this also acts as an advertisement for Scotland: "where history still lives!" This model of a dual-speed historical conveyor belt—in which Scotland has recently stepped onto a particularly fast track, transitioning from the hunting and pastoral stages through agricultural to commercial in a mere sixty years—proved impressively tenacious. Even half a century later, we can still see the civil servant and historian Spencer Walpole (1839–1907) imbibing and reinforcing Scott in his *History of England from the Conclusion of the Great War in 1815* (1876–1884). Walpole declares that:

> Scotland, in one sense, was a much more backward country than England. In the middle of the eighteenth century, English roads were intolerably bad; but Scotland, it might almost have been said, had no roads. English agriculture was backward; but Scotland was uncultivated. . . . A journey from London to Edinburgh was a more difficult and a more hazardous undertaking than a journey from London to New York is now. . . . Yet the development of Scotland was proceeding at least as rapidly as that of England and Wales. . . . In the latter half of the eighteenth century, Scotland made unexpected progress. Her lowlands were gradually converted from a barren waste into the garden of Great Britain.[29]

Scott's self-promoting rhetoric had found a place in the magnum opus of a historian and "liberal codifier" who helped to cement the accepted narrative of nineteenth-century history.[30] The comparison of Scotland with the Elizabethan era has been translated into the geographical dimension, but this comparative analogue clearly had both a long life and a long reach.

The elision of geography and time serving for one another is the enabling construct of the comparative method. Mikhail Bakhtin describes Scott as able to "read time in space," and Franco Moretti specifies this as relating particularly to "internal border[s]," the spatial analogues of times of transition.[31] Scott amplifies what was already a common comparison of Scottish pasts with contemporary primitive cultures, especially Native Americans.[32] He even suggests that failure to recognize the truth of this model can be literally dangerous. In *Waverley*, we see this in the preliminary musings of our hero's father, Richard (the apostate Whig to his brother Sir Evarard's high Toryism), as he deliberates on how best to separate Edward from an unsuitable girl at home: to send him abroad or into military service? The young man's tutor advocates the army as a safer option than a continental Grand Tour, arguing, "What might Mr Edward Waverley's society be at Paris, what at Rome, where all manner of snares were spread by the Pretender and his sons; these were points for Mr Waverley to consider. This he could himself say, that he knew his majesty had just such a sense of Mr Richard Waverley's merits, that if his son adopted the army for a few years, a troop, he believed, might be reckoned upon in one of the dragoon regiments lately returned from Flanders."[33] This logic implies that service in the British army is the safe option; but it is forgetful of the "Other" within.

As the reader sees before long, exactly those kinds of dangers are present very close at hand. Once Waverley reaches the Lowlands of Scotland and a new love interest, Rose, tells him of the horrors she has seen when a "skirmish" with the Highlands came perilously close to her house, "Waverley could not help starting at a story which bore so much resemblance to one of his own day-dreams. Here was a girl scarce seventeen . . . who had witnessed with her own eyes such a scene as he had used to conjure up in his imagination, as only occurring in ancient times." As Scott describes it, "It seemed like a dream to Waverley that these deeds of violence should be familiar to men's minds, and currently talked of, as falling within the common order of things, and happening daily in the immediate neighborhood, without his having crossed the seas, and while he was yet in the otherwise well-ordered island of Great Britain."[34]

Comparative history is at work here. Waverley assumes (in line with his teachings) that foreign countries are more backward, so has he just stepped into one of those? Part of the excitement of *Waverley*, to its first readers, lay in its exposure of *unheimlich* horrors, which English society thought it had banished to the past and overseas but which were actually still within what Moretti visualizes as "internal borders." This could be all the more easily translated into contemporary concerns as the century went on and the governmental borders of the British Empire expanded to include indigenous cultures that seemed, in

Edward Burnett Tylor's notion, like "survivals" from an earlier stage of human history.[35] Even some of these colonial outposts, however, might still be seen as uneasily internal. As Catherine Hall describes Jamaica's place in the British imaginary, it was an "outpost of the metropolis, an extension, or perhaps excrescence of the self . . . inside rather than outside": the implication being that it could "be milked for all its worth with no returns," a policy that of course led to backlash and rebellion in 1831 and 1865.[36]

Commentators recognized and endorsed this use of space as time and vice versa. William Hazlitt (1778–1830) in an assessment of his key contemporaries, *The Spirit of the Age* (1825)—whose nominal "great men" framework is tempered by a Romantic concept of a zeitgeist with an agency of its own—asks rhetorically, "Have we no materials for romance in England? Must we look to Scotland for a supply of whatever is original and striking in this kind?" His answer is "'Yes!' Every foot of soil is with us worked up: nearly every movement of the social machine is calculable. We have no room left for violent catastrophes, for grotesque quaintness, for wizard spells." His proof for this is that "even Sir Walter is ordinarily obliged to pitch his angle (strong as the hook is) *a hundred miles to the North of the 'Modern Athens'* [Edinburgh] *or a century back*."[37] In this formulation again, geographical and temporal distance are interchangeable.

We see this common trope in play throughout the first half of the nineteenth century. Returning to his discussion of historiography, Macaulay concludes that "the effect of historical reading is analogous, in many respects, to that produced by foreign travel. The student, like the tourist, is transported into a new state of society."[38] Here he is surely thinking particularly of Scott's novels, which he perceived as such competitors with conventional history in the 1820s and 1830s, and which fueled his aspirations for an equally evocative historiography. And this trope is by no means confined to historians. The novelist Elizabeth Gaskell (1810–1865) muses, at the opening to a short story set in Wales, that "Welsh people . . . are what I suppose we English were a century ago."[39] For the English, the Welsh had been the "Other" from the thirteenth century through to the sixteenth, from Edward I's conquests to Wales's final assimilation under Henry VIII. Eighteenth-century Britain's "Other" within was Scotland, particularly its Highland, Catholic, and Jacobite half. The nineteenth-century equivalent was Ireland. Its place in the British imagination, however, was complicated by the parallel concern of the ever-expanding British Empire overseas. The potential racialist parallels this opened up come into focus now in our next case study, Thomas Carlyle's "An Occasional Discourse on the Negro Question" (1849), which he wrote after a trip to Ireland in the wake of the Great Famine.

THOMAS CARLYLE AND JAMAICA

The gradualist model of history we discussed earlier not only denied the clear and momentous watersheds between stages so dear to catastrophist models. It also placed sharp limitations on available models of historical agency. As Burrow describes it, Victorian historians were faced with the problem of "how one speaks of a history without heroes, almost without events, a history essentially of largely anonymous agents and unintended consequences," a problem only exacerbated by Darwinian theory's view of change as an incremental process removed from direct agency.[40] We can see this dilemma epitomized in the work of that prolific writer on historical agency, Thomas Carlyle. In his early work, such as the anonymous 1830 article "On History," he challenges accepted norms of what determines historical significance. "When the oak-tree is felled, the whole forest echoes with it; but a hundred acorns are planted silently by some unnoticed breeze. Battles and war-tumults . . . are remembered by accident, not by desert."[41] In contrast to these sentiments, Carlyle is now often best known for his "Heroes and Hero-Worship," a preoccupation later reinforced by his biographies of Oliver Cromwell (1599–1658) and Frederick the Great (1712–1786). As David Amigoni characterizes it, Carlyle's legacy is "open to multiple cultural appropriations."[42] Despite a shift in his late career toward a preoccupation with more conventional and conservative hero figures, Carlyle remained open to appropriation as the champion of the ordinary and insignificant individual.[43] His dialectical insistence on the need to see beyond obvious heroes in tandem with his irrepressible desire for heroism can be seen as a response to this sense of a sedimentary history of anonymity. Dominated by the sense of a history "without heroes," Victorian writers became obsessed with the question of the relationship between the individual and history and with the extent to which any individual can influence their surroundings.

We can see the latter in the work of Galton, but the former preoccupation is epitomized in Carlyle.

Carlyle's "Occasional Discourse on the Negro Question," imbued with the rhetoric of the comparative method, was published anonymously in *Fraser's Magazine* in 1849. It was later expanded and reissued in pamphlet form in 1853 under a new title where "Negro" became "Nigger." This extended edition, in which Carlyle's racialist discourse is most explicit, is the one I analyze here. It drops the bracketing construct of a found text it acquired for *Fraser's Magazine* but retains the narratorial persona of a less than universally admired "Speaker" addressing a philanthropic society.[44] The tone and frame of reference are set in this passage from near the opening of the essay:

> How pleasant, in the universal bankruptcy abroad, and dun, dreary stagnancy at home, as if for England too there remained nothing but to suppress Chartist riots, banish united Irishmen, vote the supplies, and *wait* with arms crossed till black anarchy and social death devoured us also, as it has the others; how pleasant to have always this fact to fall back upon: our beautiful black darlings are at least happy; with little labor except to the teeth, which, surely, in those excellent horse-jaws of theirs, will not fail! . . . beautiful blacks sitting there up to the ears in pumpkins, and doleful whites sitting here without potatoes to eat: never till now, I think, did the sun look down on such a jumble of human nonsenses.[45]

As Catherine Hall describes it, this "marked the moment when it became legitimate for public men to profess a belief in the essential inferiority of Africans, and to claim that they were born to be mastered and could never attain the level of European civilization."[46]

The force and focus of Carlyle's racial comparison is clear from this opening. Catherine Gallagher has shown how common was this juxtaposition of workers and slaves (or former slaves) to rail against the latter's superior living conditions.[47] It was used throughout the first half of the nineteenth century and in campaigns for industrial reform as well as in pro- and antiabolition arguments. In the service of the latter campaign, Carlyle depicts the freed slaves of Demerara as better off than the poor of industrial Britain. He presents freed slaves using the familiar orientalizing tropes of "idleness" and evokes the fearsome figure of the socially elevated "idle black gentleman," an uncanny return of the repressed.[48] And what is the cause of this "rotten mass of idleness"? "Strong sun supplies itself gratis, rich soil in those unpeopled, or half-peopled regions almost gratis; these are his 'supply,' and half an hour a day, directed upon these, will produce pumpkin, which is his 'demand.'"[49] The climate theory of race, which was codified by the 1770s and which runs right through to Max Weber's *The Protestant Ethic and the Spirit of Capitalism* (1905), is heavily in evidence here. It is combined, though, with the idea soon to be popularized by Robert Knox that, rather than climate having the power to transform the characters of races, the "negro character" was irrevocably limited by the climate to which it was molded.[50] Carlyle's famous mockery in this article of economics as "the dismal science" comes partly from this conundrum: his belief that the freed slaves of the Caribbean have no deeper motives for work than the physical compulsion they were accustomed to as slaves, so that the lauded dictates of supply and demand will not function in such a tropical climate and alien mentality.

The philosopher Mary Midgley has described what she calls "the Paradox of One-way Equality": the impossibility of distributing "equality" universally.[51] She highlights the tendency for groups campaigning for equality to be blind to—or even to disregard deliberately—the claims of groups even further down the pecking order. This is perhaps at the heart of the analogy making of comparative history. Locating another culture, civilization, or race at a temporal as well as geographical remove from one's own, as we have seen, can be fitted into a monogenist conception of the human species: they might never catch up with us, but they will reach our present in the proper parallel timescale. It can also, however (as we see here), enable a polygenist conception, enabling one to categorize it as incommensurably different, eternally separate: in a different historical trajectory. If the "quashee" is naturally and inevitably idle, Carlyle's rationale goes, then he *cannot* become like us, cannot follow the Carlylean philosophy of work. Then, as he argues in a passage added for the 1853 pamphlet edition, we are justified in bringing him back into a condition of slavery or serfdom, for his own good and for the good of the bizarrely bathetic "sugars, cinnamons, and nobler products of the West India Islands, for the benefit of all mankind."[52]

We can see exactly this same unfortunate deployment of the "Paradox of One-way Equality" at work in the contemporaneous discussions of Chartist activity. The poetry of Chartist activists draws repeated parallels between their plight and that of African slaves. They use this comparative discourse to imagine an international brotherhood of dehumanized laborers. At times, however, they turn the comparison into a competition. What marks Chartist and antislavery poetry apart from Carlylean discourse is its shared internationalist subversion of patriotic ideals. In the perpetual historiographical puzzle of why the Chartists failed, one simple answer lies in their failure to gain the ear of Parliament, despite their mammoth petitions. They mainly existed outside the "political nation," and in the same way, their poems step outside conventional nineteenth-century British patriotism. Gerald Massey's "Song of the Red Republican" (1850) declares, "We'll tread them down yet curse and crown, Czar, Kaiser, King, and Slave," in a juxtaposition that suggests these titles are merely different names for the same tyrannical power.[53] His faith in universal international congruence is made explicit in Massey's "Kings are but giants because we kneel," which proclaims:

> We'll never fight again, boys, with Yankee, Pole and Russ,
> We love the French as brothers, and Frenchmen, too, love us![54]

This internationalism invoked the disgust of Caroline Norton (1808–1877), whose 1848 "Letters to the Mob" accuses the Chartists of being "so transformed into monkeys and parrots, that you raise the banner and shout the motto of a foreign people in our streets, as if you had no country and no language of your own."[55] The polygenist "monkeys and parrots" suggests that in forgoing patriotic duty for internationalism, Chartists have degraded into—or revealed their true—subhuman forms.

For both Chartists and their antagonists, the relative importance of abolitionism in relation to campaigns for the industrial poor is perpetually a point of contention. Hall relates how Maria, Lady Nugent (1814–1883), Irish wife to the 1801–1805 governor of Jamaica, saw the slaves' living conditions as better than those of her fellow Irish.[56] This frustration does find echoes in Chartist poetry. An 1840 poem in the *Chartist Circular* by "A.M.P.," satirically titled "The Land of Freedom," comes uncomfortably close to Carlyle's vision of "the regions of slavery" in implying that the freedoms of an industrial worker are the more cruelly circumscribed. Many Chartist poems highlight the plight of slaves parted from their children.[57] But the shock of "The Land of Freedom" lies in its bringing this common trope home to the "free ground" of England in a Carlylean return of the repressed. In this poem, Mohab the slave sees a British man "with sunk eyes and parch'd tongue" reduced to the poorhouse, where "they from him his wife and his little ones tore, / And to separate dungeons conveyed," with the anonymous pronoun "they" eliding all sources of cruelty and oppression into one shadowy force.[58] This disregard for the family unit supposedly hallowed by British culture suggests that industrial workers are not considered British citizens: instead, their true status is that of slaves.

Carlyle's repeated evocation of the former slaves of the Caribbean as a horrific kind of "black Ireland" draws our attention to how pervasively these different "Others" of white patriarchal British culture are elided.[59] Midgley highlights that the "four distinct problems" of "the position of women, of slaves, of other races, and of non-human animals" need to be considered together, "not because their logic is necessarily similar, but because their history is so"—and I would add, because the comparative method so often brings the first three into conjunction (and after 1859, the fourth as well).[60] Carlyle wrote his "Occasional Discourse" in the wake of a trip to Ireland, in preparation for an intended book on the "Irish problem," but as Julie Dugger puts it, "the condition of Ireland [was] unthinkable," making slavery "thinkable" instead.[61] Carlyle struggles in this piece between positing black/white binaries and eliding black with Irish: he wrote in his reminiscences of the trip that the "true station" of the Irish "is *servants* . . . 'slaves' if you will."[62] Dugger highlights that Carlyle wrote in these terms about Ireland partly "to resist progressive narratives depicting that country's plight as a developmental stage to be suffered through rather than repaired."[63] It is his way of resisting a progressionist stadial teleology, but it backfires into primitivism.

In his earlier and more radical days in *Past and Present*, Carlyle had described a "poor Irish widow" who was refused help by all "Charitable Establishments" of Edinburgh, and who eventually

> sank down in typhus-fever; died, and infected her Lane with fever, so that "seventeen other persons" died of fever there in consequence. The humane Physician asks thereupon, as with a heart too full for speaking, Would it not have been *economy* to help this poor Widow? She took typhus-fever, and killed seventeen of you!–Very curious. The forlorn Irish Widow applies to her fellow-creatures,

> as if saying, "Behold I am sinking, bare of help: ye must help me! I am your sister, bone of your bone; one God made us: ye must help me!" They answer, "No; impossible: thou art no sister of ours." But she proves her sisterhood; her typhus-fever kills *them*: they actually were her brothers, though denying it! Had man ever to go lower for a proof?[64]

Here Carlyle evokes the monogenism of the abolitionist motto "Am I not a man and a brother?" to argue for mutual dependency: even the "poor Irish widow" has the agency (even if advertently) to determine the life and death of her fellows. By 1849 the threats within Britain's "internal borders"—the specter of Chartism and Irish rebellion—has made him less tolerant both of those internal forms of dissent and of demands from anywhere else in the empire.

FRANCIS GALTON AND SCOTLAND AGAIN

As we have seen, a stadial model, and particularly the comparative rhetoric it enables, was used for both unifying and divisive purposes. In the latter half of the nineteenth century, however, it came to be used in diverse alternative formulations, with the lines of division between races being forced down invisible—or imagined—hairline cracks in society. What did this mean for so-called survivals? To answer this question, let us turn to our third and final case study, Francis Galton, and his famous work *Hereditary Genius* (1869). Grandson of the eminent Erasmus Darwin (1731–1802), and cousin of Charles Darwin (1809–1882), Galton was obsessed with hereditary intelligence. As Ruth Schwartz Cowan shows, this was despite (or because of) the fact that, though a "child prodigy" as a toddler (according to his siblings), he did not do well either at school or at university. In *Hereditary Genius* he turns a blind eye to any kinds of inheritance other than the (genetic) transmission of intelligence—including wealth, opportunity, and what we would now call cultural capital—even though at the age of twenty-two he inherited enough of his father's fortune that he never had to work for his livelihood. His life and work is thus full of ironies and, we might say, self-delusions.[65]

The idea (as Tylor would soon express it) that "survivals" from earlier stages should not be nurtured or treasured but, instead, expunged to make way for the next stage is prominent in Galton's work. This is rooted in a progressionist as well as a comparative model of history. Only at the very end of his book does he explicitly tie his work to the question of human origins, but here he pulls no punches in his attack on biblical and Adamite narratives of human history, to declare that far from having "fallen from a high estate," "the human race were utter savages in the beginning," and have only recently risen to "morality and civilization."[66] As Emel Aileen Gökyiğit has shown, the insistent "naturalism" of this self-professed agnostic earned him condemnation by reviewers in Christian periodicals but approval from the growing community of scientific naturalists.[67] He is also very much of the "great men" school of history. If Scott dealt with the transition between stages by having his protagonist Waverley "waver" between Jacobite and Hanoverian, and Carlyle dealt with the clash of sedimentary and catastrophist models by wavering between dispersing agency among "a hundred acorns" or the French Revolutionary crowd and calling for "heroes" to be worshipped, Galton dealt with the lack of agency inherent in Darwin's ideas of random mutation and unpredictable natural selection by reverting to that latter Carlylean model. This fuels all his research and conclusions: "If we could raise the average standard of our race only one grade, what vast changes would be produced! . . . We know how intimately the course of events is dependent on the thoughts of a few illustrious men."[68]

Galton thus distorts and subverts Darwin's ideas, even while he vaunts his Darwinian credentials. In *Hereditary Genius*, as he transfers his attentions from "the kinships of individuals" to "a consideration of nations and races," he explains adaptation as "owing to the sure operation of Darwin's law of natural selection."[69] He takes the ideas of natural selection that Darwin applied to species, however, and subdivides his groups to a biologically unfeasible extreme. For Galton, different regional groups can be seen as different races, and he yet again reverts to Scotland as an example, to declare, "The average standard of the Lowland Scotch and the English North-country men is decidedly a fraction of a grade superior to that of the ordinary English, because the number of the former who attain to eminence is far greater than the proportionate number of their race would have led us to expect."[70] This only works because for Galton, even higher and lower social classes are classified as stronger and weaker races. In France, for example, the revolutionary guillotining of the nobility has "made sad havoc among the progeny of her abler races."[71] The word becomes little more than a tool of social segregation, but one that Galton sees as obvious, natural, and unproblematic.

As a result, Galton tries to use "scientific methods," and especially the bell curve of normal distribution, to pin down something malleable—while dismissing the significance of external circumstance, context, and culture. As he declares in his preface to the revised edition of 1892, he explicitly excludes consideration of "the effects of education," trying to evaluate only "natural ability."[72] However, (one of) the largest unacknowledged assumption(s) in his argument is that there is a mutual and equivalent two-way causation between eminence and intelligence. Because, in his formulation, "intelligent" can be "expressed by the word 'eminent,'" eminence is caused (primarily and even solely, in an exclusive and scientifically measurable sense) by intelligence![73] Galton relates how he has convinced many "incredulous friends" of his once they have realized how many eminent men "have eminent relations."[74] As many reviewers protested, this could result from "fortune" and "favors," or as *Chamber's Journal* summed it up, "a helping hand."[75] The extent of this pseudomathematical delusion is highlighted by the largest absence from his data: women. Although the *Dictionary of Men of the Time* that he uses actually includes "eminent living characters of both sexes," the notion of eminent women is simply ignored by Galton.[76] Women appear only as maternal ancestors. Even then, in his quantitative tabulations, he signifies "mother" with a "small italicized" *f* (derived from the code for "father," capital *F*), blind to the possibility of using *P* for parent.[77]

As all this demonstrates, Galton's insistently scientific data is shaped by ideology. His tabulated grades of intelligence are "pure conjecture," as Sherrin Berezowsky puts it. "Galton had no data whatsoever to populate [his] table," but it drives how he reads the data that does arise.[78] Occasionally, as in his later work *Fingerprints* (1892), he admits the need to subordinate ideology to the lack of significant results, relating that "it seemed reasonable to expect to find racial differences in finger marks," but that the fingerprints of all the races he examined "may all be spoken of as identical."[79] Despite all evidence to the contrary, however, he adds that "still . . . [even if] pure fancy on my part . . . the general aspect of the Negro print strikes me as characteristic."[80] As Colin Kidd highlights, this is fallacious, and the map of apparent affinities from fingerprints would link disparate geographical areas and "confound . . . our expectations of racial classification."[81] Galton's aspirations for his fingerprint studies are to find hierarchy as well as for differences: "the Hill tribes of India," for example, he thinks might show a "more monkey-like pattern."[82]

The ideology that drives him, therefore, is one of racial hierarchy. He evokes the same hierarchy, in the same (Darwinian) animal terms, from the opening page of *Hereditary Genius*. "As it is easy . . . to obtain by careful selection a permanent breed of dogs or horses gifted with peculiar powers of running, or of doing anything else, so it would be quite practicable to produce a highly-gifted race of men by judicious marriages during several consecutive generations."[83] It might not be the animal imagery that seems most ominous here, but—as we have seen in Carlyle's evocation of the "negroes'" "horse-jaws" and "beautiful muzzles," "bleating and braying"—it is racially motivated.[84] Galton's animal imagery reflects onto other human "races" and sees them in a comparative and stadial mold, looking backward as well as forward. In the preliminary manifestation of his ideas, his 1865 articles in *Macmillan's Magazine* on "Hereditary Talent and Character," he concludes: "Men and women of the present day are, to those we might hope to bring into existence, what the pariah dogs of the streets of an Eastern town are to our own highly-bred varieties."[85] Here, Galton's comparative approach cuts across species, racial, and stadial boundaries. He simultaneously compares backward from British pedigree dogs to supposedly less-advanced "pariah dogs" of (we can assume) the South Asian colonies and proposes the resulting measurement of differential progress as equivalent to one between current British people and those of a utopian eugenic future. This comparative model is a notably *flat* one, where anything and everything is comparable since it all—species, race, IQ—follows the same evolutionary model in macro- or microcosm. Bowler has suggested that the "darker side" of the progressionist model of history is the inevitable approach of "senility and death," but the "darker side" might more aptly be named as the implicit—and explicit—value judgments placed on these different cultures that had progressed different distances up the scale.[86]

It is worth acknowledging that Galton's comparative method results from feeling daunted. He is driven in part by a belief that modern life is more complex than any preceding form, and thus that it requires a higher level of intelligence. Otherwise, as he perceives it, "our race . . . appears likely to be drudged into degeneracy by demands that exceed its powers," namely, "the level of work required by the nineteenth century."[87] He plans a eugenic world—one where people of high and low "intelligence" would voluntarily hasten or delay procreation for the good of society—in order to live up to the juggernaut of material progress. This is fundamentally a question of agency: Who gets to speed up or slow down progress? Who or what takes a society over the watershed into the next stage? This element of Galton's work is often unacknowledged. Among studies that condemn him as a proponent of unethical and destructive policies and those that try to defend his record as biometrist, and thus alternately demonize or champion him, few acknowledge that his declaration of agency was at root a response to a perceived *lack* of agency.[88] This returns us to the conundrum persistently tracked and reshaped by Hegel, Comte, and Marx: What more vivid epitome exists of the tension between individual agency and the impersonal forces of history than the predictive, prescriptive, but also intentionally empowering dialectics and teleologies of these programmatic thinkers?

As we have seen, therefore, the stadial model of history—and the comparative method it enabled—was surprisingly resilient through the nineteenth century, but it shifted from conjectural to more hostile and polygenist forms from the mid-century onward. It enabled each of our three writers to deal with their particular anxieties in different ways. For Scott, the stadial model enabled him to explain the difference between Scotland and England, to make his novels believable, and to advertise their version of Scotland as a place of sublime escape. For Carlyle, who was famous as a prophet of nonprogress and who in *Past and Present* had mocked the current "godless century" for failing to appreciate previous "centuries that were godly," stadial progression is not a given, nor is the equal applicability of this process to all civilizations and races. On the contrary, the "white man" has a sacred duty of work, whereas the "black man" is fundamentally idle, such that the Caribbean would have remained an uncultivated swamp "for ever" if colonizers had not rescued it. Thus, while he rejects some facets of the stadial model, he retains its transformational drama and its comparative leverage.[89] For Galton, the transformative watersheds have had to be put aside in the wake of his cousin's gradualist theory of evolution, but (if one intervenes to subvert the nondirectionality of Darwin's thesis) what is still usable is the progressionist trajectory, and the upward—antibiblical—narrative from savagery to civilization. In Galton, it is the comparative aspect that comes to the fore and is elaborated to an unprecedented (and to us ludicrous) extent, as upper and lower classes, Scottish and English, are constructed as separate and potentially divergent races.

Why were these stadial models of history so pervasive and persistent? One reason might be that, unlike the sedimentary gradualist model, their teleological and universalizing claims enabled overview from within. They allowed their users to feel like they had access to a broad map on which all the civilizations they encountered could be placed. They gave immediate clarity—instant gratification, one might say—and offered up the potential to predict the future. However, these methods of comparative and conjectural history, and the consequent notion that some cultures and civilizations were closer to the origins of "man" than others, were as limiting as they were enabling. As Burrow put it half a century ago: "To be able to describe a puzzling social phenomenon with comfortable finality as a 'survival' converted into an intellectual resting-place what might otherwise have been a starting-point."[90] By providing more or less comfortable answers to vexing questions, such description blocked the potential for further thought. It risks the same kind of fixity as the age-old Great Chain of Being.

8

THE FUTURE EVOLUTION OF "MAN"

IAN HESKETH

When Charles Darwin (1809–1882) finally applied his evolutionary theory of natural selection to humans in the *Descent of Man* (1871), a full twelve years after writing *On the Origin of Species* (1859), he stressed the scientific nature of his enterprise: "We are not here concerned with hopes or fears," Darwin argued on the final page, "only with the truth as far as our reason allows us to discover it."[1] This followed directly from his argument that, on the basis of the preceding evolutionary narrative, humans may look forward to a "still higher destiny in the distant future," which suggests that hopes and fears were actually more central to establishing the natural history of humans than Darwin was willing to admit. Moreover, his appeal to a "distant future" was precisely what evolutionary theorists and writers did to compensate for the fact that humanity's origins were now deemed to be subject to the same evolutionary processes that gave shape to all other forms of organic life.[2]

There was, therefore, a necessary tension in establishing a natural evolutionary history of humans, because doing so challenged explicitly the special status of humanity vis-à-vis the rest of the natural world. And Darwin's theory of evolution by natural selection seemed particularly harsh and devoid of direction or guiding purpose such that the arrival of humans could be interpreted as a mere accident of the blind laws of nature. This is what the astronomer John Frederick William Herschel (1792–1871) referred to when he famously said that natural selection was the "law of the higgledy piggledy."[3] Moreover, that humans might be descended from a beast was another significant blow to humanity's "dignity" and indeed related to one of the main fears about Darwinian evolution—namely, that it was an essentially meaningless process. It was precisely the nonteleological nature of Darwinian evolution, according to Peter Bowler, that led the British public, as well as the vast majority of naturalists, to reject natural selection as the primary mechanism of evolutionary change.[4] But this did not necessarily mean that, because of their religious worldviews, they rejected evolution in toto.

Historians of science have long stopped viewing the establishment of evolution in the nineteenth century as evidence for a larger "conflict thesis," which posited that science and religion have been in a state of constant conflict, a thesis that was itself a construct of the period.[5] It must be said that there certainly was conflict between contending groups vying for cultural authority, such as an older generation of gentlemanly naturalists and a younger generation of scientific naturalists, who immediately embraced evolutionary science as supporting their professional endeavors.[6] Darwin provided evolution with a certain amount of scientific legitimacy, but the theory itself proved highly malleable, as even high-profile scientific naturalists such as Thomas Huxley (1825–1895) and Herbert Spencer (1820–1903) embraced competing evolutionary theories.[7] Moreover, it is now well established that the general idea of evolution achieved fairly widespread consensus, in part, thanks to the ubiquitous work of popularizers of science, who often made evolution more palatable to changing Victorian religious sensibilities.[8] But how was this possible, if a theory of evolution necessarily posited a natural origin for humanity?

Part of the explanation must be sought in the fact that, although Darwin was at least somewhat hesitant to indulge the hopes and fears of his readers, many other evolutionary theorists and writers were not. These other writers did so by explicitly discussing the larger meaning of the evolutionary process, particularly in regard to a now naturalized human origin, within the framework of a grand narrative of life. In accepting that Darwin had at the very least undermined the notion of a supernatural origin of humans, these other writers sought to replace that myth with a naturalized one. Doing so entailed seeking the mythological meaning of evolution, not in the past, but in the distant future where adequate compensation could be sought for the perceived loss of purpose that was entailed in human evolution. And, perhaps unsurprisingly, such a futuristic vision necessarily infused the evolutionary process with a meaning that was decidedly metaphysical, if not religious, no matter how secular it actually claimed to be.

So what role did the future play in Victorian evolutionary narratives that sought to situate humans within a larger evolutionary history of life? In this chapter we will survey evolutionary writers from the 1860s until the turn of the twentieth century, most notably Alfred Russel Wallace (1823–1913), W. Winwood Reade (1838–1875), E. Ray Lankester (1847–1929), Francis Galton (1822–1911), Frederic Myers (1843–1901), and John Page Hopps (1834–1911).[9] While some of this material will be familiar to historians of evolution, the focus on the futuristic visions of these evolutionary narratives will allow new light to be shed on the metaphysical issues their authors necessarily confronted at the end of the nineteenth century. Even though the religious views of these evolutionists ranged widely from the agnostic to the spiritual, they all traded on notions of Christian eschatology in order to imbue their evolutionary narratives with a grand cosmic meaning.

How did these evolutionists seek to project their naturalized interpretation of humanity's evolution into the future? Contemporary discoveries in biology, physics, and cosmology necessarily problematized such teleological narratives of eternal progress. Indeed, toward the end of the nineteenth century, burgeoning scientific theories such as degeneration and entropy destabilized the future as a space, whereby earthly utopias could be projected as the natural outcome of the evolutionary process. The notion of evolutionary degeneration coupled with the inevitability of heat death ramped up the stakes for the future evolution of humanity, which was now deemed a process that could not be left to nature alone. Moreover, it was at this time that speculation about humanity's non-terrene evolution began, as it was postulated that the Earth was simply a first stage in what would likely be a lengthy evolutionary future. In this regard, speculations from both eugenics and spiritualist theories of evolution confronted the same problem with a similar metaphysical solution concerning humanity's higher evolutionary future. By the end of the century, evolution had become a truly cosmic process. Human origins may have become naturalized, but their future promised to be one that united them with the eternal cosmic mind. While Darwin may have made a convincing case for an evolutionary history of humanity, it is clear that, in order to imbue the evolutionary process with a cosmic meaning, leading evolutionary thinkers often found themselves needing to go well beyond Darwin's work in order to do so. And rather than being the primary mechanism of evolutionary change, natural selection was ultimately reimagined as the final natural barrier to be overcome in order to achieve a truly transcendental evolutionary future.

THE NATURAL HISTORY OF "MAN"

It is well known that human evolutionary origins were very much on the forefront of Darwin's mind throughout his early inquiries into variation by natural selection.[10] However, with the exception of the rather remarkable phrase that his theory promised to "shed light on the origins of man and his history," Darwin in the *Origin of Species* was silent on how his scheme applied to humans. Why he avoided discussing human evolution has been a subject of much debate, though it seems most likely that he wanted to establish his theory of natural selection in general before getting mired in a debate about human origins. Moreover, some of his close friends such as the geologist Charles Lyell (1797–1875), were still unwilling to go "the whole orang," even while being generally convinced by Darwin about evolution concerning the nonhuman world. For Lyell and many others, the arrival of humans represented something quite new in the historical record and therefore could not be explained by the same mechanism of evolutionary change that was being applied to other organic life.[11] The fact remained for Victorian society at large that humans were special. Presenting a transformed natural world picture was one thing; explicitly including humans within that transformed scheme was quite another indeed.[12]

While Darwin would eventually write about human variation in the context of a larger discussion of animal evolution in the *Descent of Man*, a substantial two-volume work that largely focused on the mechanism of sexual selection in order to explain both racial and gender differences, for more than a decade he ceded ground on the issue to others, who were not as hesitant to apply his theory of natural selection to human origins. Of course, natural selection was "discovered" not just by Darwin but also by Alfred Russel Wallace, who is often described as its codiscoverer thanks to a timely letter he sent to Darwin outlining such a theory that forced the ever-delaying Darwin finally to publish. Wallace was one of the first to make a sustained effort to think about how the racial and intellectual distinctions found in contemporary human societies could be explained with reference to humanity's evolutionary history, particularly with reference to the theory of natural selection that Wallace himself had already done so much to help develop.[13]

In a paper titled "The Origins of the Human Races and the Antiquity of Man Deduced from the 'Theory of Natural Selection,'" which he delivered at the Anthropological Society of London (ASL) in 1864, Wallace ostensibly sought to bridge the monogenist and polygenist positions by arguing that, although humans emerged from primates as a unified species, that species subsequently split into a variety of related species largely represented by distinct racial characteristics. These physical distinctions then led to distinct mental characteristics, which have all been shaped by the "irresistible action of 'natural selection.'"[14]

At their origins, Wallace claimed, humans, much like any other species, must have been subject to the processes of natural selection. We know this is true because it is a well-proven fact that the rest of the natural world, stretching back in time to remote ages, has been subject to the same "irresistible action," with an endless array of species being formed and shaped in order to "preserve their harmony with the ever changing universe." Our understanding of the history of life, therefore, shows that "no living thing could escape the law of its being; none could remain unchanged and live, amid the universal change around it." But, Wallace explained, while humans originated much like any other species, they became quite unlike any other because of "a subtle force we term mind," which came into existence early in human evolutionary history. This was a new and entirely unique characteristic that became much more important than the physical structure in which it was contained. What Wallace therefore hoped to show was that his argument about the evolution of the human mind set "man apart, as not only the head and culminating point of the grand series of organic nature but as in some degree a new and distinct order of being."[15] Wallace stressed that he was not so much bringing humans down to the level of the ape, as was the seemingly logical conclusion of the theory of evolution by natural selection, but rather, using the theory in relation to the development of mind to suggest that humans were indeed a "higher" species.

Wallace stressed that the theory of natural selection could account entirely for the origins of the human species. However, once natural selection worked to establish the human mind, the mind quickly evolved, thereby allowing humans to use nature to their advantage, creating clothing and weaponry, and even controlling the growth of food through planting and domestication. By understanding the way the human mind has evolved over time, argued Wallace, we gain insight into the fact that humans eventually came to dominate and control nature in a way that was simply impossible for other species. It is by focusing on the "advance of mind," argued Wallace, that we "see the true grandeur and dignity of man. . . . He is, indeed, a being apart, since he is not influenced by the great laws which irresistibly modify all other organic beings."[16]

Wallace, therefore, believed he proved a seemingly contradictory thesis that although humans were the products of natural processes they evolved their intellectual capabilities to transcend those very same processes. And he went still further. "Man has not only escaped 'natural selection' himself, but he actually is able to take away some of that power from nature which, before his appearance, she universally exercised."[17] What Wallace meant by this was that humans had evolved to such an extent that it was now possible for them to take over the processes that led to their birth in the first place and to establish a truly remarkable future existence.

Wallace envisioned "man's selection" in the future as replacing "natural selection," which is a wonderful reversal of Darwin's analogy between domestic selection and natural selection that is central to the *Origin of Species*. Whereas Darwin argued in the *Origin of Species* that humanity's humble attempts at selective breeding and domestication pale in comparison with the awesome power of nature over the space of millions of years, Wallace argued that in the very near future humans will remake the Earth anew, such that the Earth will only produce cultivated plants and domesticated animals.[18] He also postulated that while humanity's physical form will largely remain the same, "his mental constitution may continue to advance and improve till the world is again inhabited by a single homogenous race, no individual of which will be inferior to the noblest specimens of existing humanity." Wallace had high expectations for this future as he envisioned a utopia that conveniently dovetailed with his socialist commitments:

> Each one will then work out his own happiness in relation to that of his fellows; perfect freedom of action will be maintained, since the well balanced moral faculties will never permit any one to transgress on the equal freedom of others; restrictive laws will not be wanted, for each man will be guided by the best of laws; a thorough appreciation of the rights, and a perfect sympathy with the feelings, of all about him; compulsory government will have died away as unnecessary (for every man will know how to govern himself), and will be replaced by voluntary associations for all beneficial public purposes; the passions and animal propensities will be restrained within those limits which most conduce to happiness; and mankind will have at length discovered that it was only required of them to develop the capacities of their higher nature, in order to convert this earth, which had so long been the theatre of their unbridled passions, and the scene of unimaginable misery, into as bright a paradise as ever haunted the dreams of seer or poet.[19]

This futuristic narrative of humanity's origins was delivered as a paper at the ASL. If we are to take the comments following Wallace's paper as an indication, the ASL was largely unsympathetic to his particular attempt to naturalize the history of humans.[20] Formed in 1863 as a breakaway group from the Ethnological Society of London (ESL), the ASL was led by the physician and speech therapist, James Hunt (1833–1869), who, like most ASL members, was largely unsympathetic to Darwinian evolution, which had gained favor at the ESL. Moreover, Hunt and many of the society's members were also adherents to polygenism and tended to believe that the key facts of anthropology were discerned through comparative anatomy.[21] Wallace may have sought to bridge the gap between polygenist and monogenist views with his history of humanity, and he even argued that humans would eventually overcome the Darwinian constraints that governed the rest of the natural world, but he was ultimately relying on an interpretation of natural selection to explain humanity's history. And the members in attendance simply could not get beyond Wallace's reliance on Darwin. As Hunt himself lamented, "I think it a pity that the two subjects of Darwin's hypothesis and Mr. Wallace's paper should have been so mixed up this evening." Moreover, that Wallace ultimately argued that humans must "have sprung from one race" was deemed "to be hardly a satisfactory argument." He was, according to Hunt, basing his views on a hypothesis rather than on a clear induction of the known facts.[22]

Darwin, however, was absolutely delighted with Wallace's paper and did not suggest any concerns with Wallace's argument that humans had now taken over the processes of natural selection. "The great leading idea is quite new to me," Darwin wrote to Wallace, "viz that during late ages the mind will have been modified more than the body; yet I had got as far as to see with you that the struggle between the races of man depended entirely on intellectual & *moral* qualities." While he admitted that there were some differences between their views of human evolution, and he gestured to his theory of sexual selection that would be further explication in the *Descent of Man*, he was particularly taken with Wallace's speculative conclusions about the future of humans, believing that "the latter part of the paper" was "grand & most eloquently done."[23] Perhaps this was because some of the language seemed to be borrowed directly from the final page of the *Origin of Species*, where Darwin first used the term "evolved" with reference to the process being grand and beautiful. The difference, however, was that, whereas Darwin suggested "there is grandeur in this view of life," as understood via the mechanism of natural selection, Wallace argued that the *real* grandeur was found in human evolution, precisely because it showed that humans transcended the very natural mechanisms that gave birth to them.

If under Wallace's scheme the past history of humanity was largely determined by natural selection, human emergence from that history into a world that can be shaped to their will seems much more Lamarckian than Darwinian and shares much with the cosmic evolutionism of Herbert Spencer, and Robert Chambers's anonymous *Vestiges of the Natural History of Creation* (1844). *Vestiges* popularized a Lamarckian view of evolution as a guided process beginning in the cosmic dust and ending with the development of the human mind. There was no mechanism bringing about the evolutionary change but, rather, a force that was built into the origins of all things at the moment of creation. The arrival of humans, therefore, was presented as the inevitable result of a long process of development that might even lead to higher types of humanity.[24] Spencer believed his own evolutionary system to be much more rigorous than those of both Jean-Baptiste Lamarck (1744–1829) and the much derided (and much read) *Vestiges*. While Spencer did not adhere to the providential narrative of *Vestiges*, there was a kind of cosmic purpose to the emergence of humans as outlined in his early formulations of his scheme as the ultimate development of a long process from the general to the more specialized.[25] Wallace was, in a way, harmonizing this general developmental view with the Darwinian mechanism of natural selection. In the published version of his talk, Wallace even cited Spencer's *Social Statics* (1851) as presenting a similar view of humanity's evolutionary future.[26]

This mixing of Darwinian and Spencerian forms of evolution was not uncommon and was also produced by the anthropologist and travel writer W. Winwood Reade.[27] Much like Wallace, Reade also set out to historicize humanity within a Darwinian framework, even writing to Darwin in 1871 that his scheme would be considered a Darwinian universal history. And for the most part he was true to his word, summarizing an evolutionary history of humanity that began in the darkest corner of Africa to be extended throughout the Earth through a violent struggle for existence. But when it came time to write the section on the intellectual development of humans, Reade confessed to Darwin that he had to look beyond natural selection to explain how humans had managed to progress in a way that set them apart from the rest of nature. "I am afraid you will not find me a *good* Darwinian," Reade wrote to Darwin, "for I am inclined to believe in Natural Selection being a secondary law."[28] Natural selection, for Reade, was really only involved in the working out of physiological details, but the grand story of all life—that is, the shaping of life from the cosmic dust to the arrival and evolution of humans—was governed by a primary law Reade referred to as the "development" hypothesis.[29]

Reade's primary influence may have been Darwin, but his theory of evolution was also shaped by Wallace, Spencer, and *Vestiges*, as Reade explained in the preface to his 1872 *Martyrdom of Man*.[30] Their precise influence becomes particularly apparent in Reade's attempt to situate the intellectual development of humans within a larger narrative beginning with the origins of all life with the nebular fire mist. From the perspective of such a long developmental story of life, the incessant struggle for existence that on first glance seems to determine the evolution of the human intellect recedes from view. What appears instead is a clearly progressive and purposeful history of life that has led inexorably to the creation and evolution of humanity. Understanding this story, according to Reade, will allow humans to overcome the constraints of natural selection and bring about their final destiny: "When it is fully realized and understood that the genius of man has been developed from a long line of unbroken descent from the simple tendencies which inhabited the primeval cell, and that in its later stages this development has been assisted by the efforts of man himself, what a glorious futurity will open to the human race."[31] And for Reade this glorious futurity is one of peace, happiness, and human perfectibility that will be signified by the invention of such technological marvels as the automobile, passenger air travel, and laboratory food "in unlimited quantities at a trifling expense."[32] Much as it was for Wallace, the true meaning of evolution for Reade becomes apparent only when understood within the context of a grand narrative that envisions humans overcoming their naturalized (Darwinian) constraints in order to bring about the earthly paradise.

THE DEGENERATION OF "MAN"

When the *Descent of Man* was published in 1871, Darwin finally made public his own views about human evolution, and he proved to be much less willing to suggest that humans were somehow special or should even desire to transcend the forces of natural selection. About the former, however, his strategy was to blur the boundaries between primate and human, while in seeming contradiction to place distance between "civilized" and "savage" humans. Given that, at this point in the history of the development of evolutionary theory, most naturalists accepted that contemporary humans had developed from a previously "savage state," the difficulty was in making the case for descent from primates. In this regard, for Darwin, he argued that he "would as soon be descended from that heroic little monkey, who braved his dreaded enemy in order to save the life of his keeper; or from that old baboon, who, descending from the mountains, carried away in triumph his young comrade from a crowd of astonished dogs—as from a savage who delights to torture his enemies, offers up bloody sacrifices, practices infanticide without remorse, treats his wives like slaves, knows no decency, and is haunted by the grossest superstitions."[33] In other words, Darwin believed that "savage" humans were more violent and even more morally depraved than their primate ancestors. He argued, therefore, that his readers should have no qualms about accepting a primate ancestry.

About natural selection, Darwin sought to show that humanity's moral qualities were developed out of social instincts that were beneficial in the struggle for existence. At the same time, he was willing to accept that natural selection was no longer a primary force in "the highest part of man's nature." This is because human "moral qualities are advanced either directly or indirectly, much more through the effects of habit, the reasoning powers, instruction, religion, &c., than through natural selection."[34] Darwin stressed that this did not mean humans had somehow transcended the forces of natural selection in the most fundamental sense; that is, in regard to population pressures and reproduction. "Man, like every other animal," Darwin explained, "has no doubt advanced to his present high condition through a struggle for existence consequent on his rapid multiplication and if he is to advance still higher he must remain subject to a severe struggle." Indeed, referring to the work of his cousin Francis Galton, Darwin held out the possibility that, should humans assume they are no longer in a grand struggle for existence and fail to seek out the best mates and reproduce, "[they] would soon sink into indolence, and the more highly-gifted men would not be more successful in the battle of life than the less gifted."[35] Humanity's future evolutionary progress, therefore, was not inevitable and would not occur without engaging in the same sort of reproductive strategies that had led to humanity's current high state. This was the specific context within which Darwin suggested that, although humans could hope for a "still higher destiny in the distant future," his readers should not be concerned with hopes or fears but, rather, the truth as far as they can grasp it.[36]

Of course, Darwin *was* precisely giving voice to a set of related fears about humanity's future, fears that would only increase in the final quarter of the century. The discovery of the second law of thermodynamics in the 1850s—and the subsequent popularization of the heat death hypothesis—engendered a real fear that the eventual burning out of the sun would bring about an abrupt and tragic end to an incredible evolutionary epic of life.[37] Such a view cut against the previous assumptions of many evolutionary adherents such as Reade and Wallace that the meaning of the evolutionary process was ultimately found in the future when human perfectibility could be established. Darwin himself in his 1876 autobiography lamented "the view now held by most physicists, namely that the sun with all the planets will in time grow too cold for life. . . . Believing as I do that man in the distant future will be a far more perfect creature than he now is, it is an intolerable thought that he and all the other sentient beings are doomed to complete annihilation after such long-continued slow progress."[38] And following on from this rather gloomy future envisioned by thermodynamics, there arose a biological analogy to the law of entropy; namely, that of degeneration, which was an evolutionary concept that achieved widespread consensus in the closing decades of the nineteenth century.[39] If the general direction of the universe actually moved from complexity to simplicity, from life to death, it followed that the evolution of humans could move in much the same direction.

This was precisely the view promoted by the biologist E. Ray Lankester, who spent much of his career attempting to establish the notion of evolutionary degeneration. In *Degeneration: A Chapter in Darwinism* (1880), which was a version of his presidential address to the British Association of the Advancement of Science in 1879, Lankester stressed that the evolutionary process was not necessarily a progressive one. Central to his "Darwinian" theory of degeneration was the notion that the life cycle of an individual organism recapitulates the evolutionary development of the species as a whole.[40] What Lankester added to the recapitulation hypothesis was his argument that the life cycles of certain species actually provide evidence for evolutionary degeneration rather than progress. This was particularly apparent when one examines the development of what he called the "lower species" such as barnacles and ascidians, which arrive at their adult structures by losing many of the complex functions of their early lives.[41]

Also important to this story is what causes the degeneration. According to Lankester it is the fact that the environment of a given species seems to change in such a way to "render its food and safety very easily attained . . . just as an active healthy man sometimes degenerates when he becomes suddenly possessed of a fortune; or as Rome degenerated when possessed of the riches of the ancient world."[42] This reference to a Rome degenerated was made for analogical purposes, but only partly so. Lankester believed that degeneration was a process that was also prevalent in the history of humankind itself, a history that could be grasped by observing contemporary "savage societies," which had degenerated from previously higher states of civilization.[43] His point was that the civilized were therefore not immune to this process. Given that the "white races of Europe" had failed to progress, intellectually speaking, since the time of ancient Greece, Lankester argued that it was entirely possible that humanity was already adrift, "tending to the condition of intellectual Barnacles and Ascidians." He argued, therefore, that "it is possible for us—just as the Ascidian throws away its tail and its eye and sinks into a quiescent state of inferiority—to reject the good gift of reason with which every child is born, and to degenerate into a contented life of material enjoyment accompanied by ignorance and superstition."[44]

There was, however, a way out of this particular conundrum. And it related very much to the one way in which humans are distinct from "our ruined cousins—the degenerate Ascidians." According to Lankester, "it is possible to ascertain what will conduce to our higher development, what will favour our degeneration." This meant for Lankester that, once we understand "man's place in the order of nature," it will be possible for us to use that knowledge "to control our destinies." He argued that "we shall be able by the light of the past to guide ourselves in the future. In proportion as the whole of the past evolution of civilized man, of which we at present perceive the outlines, is assigned to its causes, we and our successors on the globe may expect to be able duly to estimate that which makes for, and that which makes against, the progress of the race."[45] Just as Reade and Wallace argued that humans had taken over the power of natural selection to shape their evolutionary future, Lankester argued that humans could avoid the coming evolutionary degeneration that his friend and colleague H. G. Wells (1866–1946) would soon after envision in *The Time Machine* (1895), but such could only be done through a sustained effort that would be thoroughly based on a sound theory of evolution.

In a later work, titled *Kingdom of Man* (1907), Lankester even more explicitly argued that, in becoming conscious of human evolution as largely inevitable and predestined from the origins of the "Kosmos," his readers had "an absolute duty" to bring about a "fulfillment of Man's destiny."[46] But Lankester further problematized this fulfillment by arguing that there was a new set of "dangers and difficulties" on the horizon, such as a possible energy shortage. Should humans be able to overcome these problems and avoid degenerating, Lankester argued that they can look to Mars, not as a future colony but, rather, as an example of what an evolved humanity might look like. The astronomical observations of the Spencerian Percival Lowell (1855–1916) concerning the possible existence of canals on Mars suggested to Lankester that this might be evidence of a highly evolved species that was able to manipulate the environment of their planet. This possibility caused Lankester to wonder if "such a future is in store for Man, that he may be able hereafter to deal with great planetary factors to his own advantage, and not only draw heat from the bowels of the earth for such purposes as are at present within his scope, but even so as to regulate, at some distant day, the climates of the Earth's surface, and the winds and the rain which seem now for ever beyond his control."[47]

The other danger Lankester envisioned was a population problem that he argued would require a thorough investigation into the "laws of breeding and heredity."[48] Because he had a different view for how to deal with this problem, Lankester failed to mention that there was already a well-known program for controlling reproduction at the time—one that was also, in theory, based on laws of heredity and breeding. This was the eugenics program of the polymath Francis Galton. It should not be surprising that Galton's work on heredity and race made explicit a vision of the future of humanity, one that would be perfected because of what he called "eugenics," or the cultivation of race. Less well known, however, is the fact that, at least in his 1883 book, *Inquiries into Human Faculty and Its Development*, Galton made the case for his program of eugenics by situating it within a grand evolutionary scheme of life, one that promised to connect the individual with the species, the specific with the universal.

Inquiries into Human Faculty built on Galton's more widely known 1869 book, *Hereditary Genius*, which postulated that it was not just physical traits that were passed down from generation to generation but mental traits as well, such as talent and character. He argued that it was possible to identify that which separated those desirable from the undesirable, the geniuses from the criminal and degenerate, and he proposed a program of early marriages for those with traits deemed most beneficial for the future of the human race.[49]

By 1883, however, fears of a degenerated future humanity were clearly taking hold, as Galton ramped up his rationale for promoting eugenics. He argued that many of his readers were largely ignorant of "the regressive direction the human race may well be headed" because so many undesirables stay out of the public sphere, out of sight and out of mind.[50]

In making the case that his program of eugenics was an answer for this problem, Galton argued that it was first and foremost "founded upon the idea of evolution." By this he meant that, once it is properly understood how humans have been shaped by evolution in the past, it will be possible to ascertain just how those evolutionary processes can be utilized to advance the cause of progress and thereby shape the human race to contemporary desires—in other words, to direct the course of evolution. In following from what he claimed was an explicitly Darwinian understanding of evolution, Galton highlighted the brutal struggle that had shaped past life, referring to the evolutionary process as an "awe inspiring spectacle of a vast eddy of organic turmoil, originating we know not how, and travelling we know not whither. It forms a continuous whole from first to last, reaching backward beyond our earliest knowledge and stretching forward as far as we think we can foresee." In this way, according to Galton, nature "has been moulded by blind and wasteful processes, namely, by an extravagant production of raw material and the ruthless rejection of all that is superfluous, through the blundering steps of trial and error."[51]

But like many other evolutionists who accepted the Darwinian struggle for existence on a local level, Galton claimed that, when we step back and consider the past from a much longer perspective, a transcendent cosmic meaning becomes apparent in the evolutionary story of life. "If we summon before our imagination in a single mighty host," Galton wrote, "the whole number of living things from the earliest date at which terrestrial life can be deemed to have probably existed, to the latest future at which we may think it can probably continue, . . . we shall plainly perceive that the actual tenantry of the world progresses in a direction that may in some sense be described as the greatest happiness of the greatest number."[52] Despite the violence and struggle that seems to be ever present on a local level, Galton argued that the larger story of evolution was a progressive one.

Now that we know this story of evolution, Galton argued, it was time for "this new animal, man . . . to awake to a fuller knowledge of his relatively great position, and begin to assume a deliberate part in furthering the great work of evolution." In order for this to happen, however, Galton was convinced that "the great work of evolution" would need to be taken up with something like religious zeal.[53] Like Lankester, Galton felt that humans needed to understand that it was their duty to take on these tasks that had been originally accomplished by the slow and violent forces of natural selection. Now that we had come to understand those processes and could deliberately, and systematically, perform them ourselves, it was our duty to "fulfill our part as members of one great family that strives as a whole towards a fuller and higher life." According to Galton, his chief goal in writing the *Inquiries into Human Faculty* was "to elicit the religious significance of the doctrine of evolution" and thereby impose a new "moral duty" and "endeavor to further evolution, especially that of the human race."[54]

He admitted, however, that if we gaze long enough into the future we will come to recognize that organic life will disappear from the face of the Earth, as the sun eventually cools and grows dark. As the conditions of existence generally worsen over time, he wrote, "there will be retrogression towards lower types, until the simplest form of life shall have wholly disappeared from the ice-bound surface. The whole living world will then have waxed and waned like an individual life." But Galton stressed that this rather depressing interpretation of the distant future of life said more about the limits to the contemporary state of knowledge than about the likely future evolution of the human race. "We as yet understand nothing of the way in which our conscious selves are related to the separate lives of the billions of cells of which the body of each of us is composed. We only know that the cells form a vast nation, some members of which are always dying and others growing to supply their places, and that the continued sequence of these multitudes of little lives has its outcome in the larger and conscious life of man as a whole." Galton, therefore, held out hope that "our part in the universe may possibly in some distant way be analogous to that of the cells in an organized body, and our personalities may be the transient but essential elements of an immortal and cosmic mind."[55] We should not, in other words, fail to accept our evolutionary duty because of the possibility that life on Earth will eventually be extinguished. There was likely a larger cosmic purpose at work in the evolutionary process that we were as yet unable to comprehend.

THE EVOLUTION OF "MAN" AFTER DEATH

Galton's hope that there may be an evolutionary future not dependent on the laws of physics, as then understood, was a point of departure for many of the evolutionists associated with the Society for Psychical Research (a society that is still in existence). Founded by Henry Sidgwick (1838–1900) and Frederic Myers in 1882, the society sought to study the largely unseen spiritual world from a scientific perspective. For some of the members, the evolutionary history of humans offered a framework for thinking about one of the most pressing issues for the society's members—namely, the possibility of a future life beyond the terrestrial one.[56]

In nineteenth-century Britain, the afterlife had essentially merged with a particularly benign view of Heaven, and many if not most Victorians held out hope that the soul transcended death into another sphere of existence. For much of the century, the afterlife was deemed largely within the terrain of Christian theology. But this began to alter substantially as the discovery of "supernormal" phenomena began to increase in the second half of the century, often during séances, where mediums purported to communicate with the dead. Such séances attracted a wide range of interest, particularly from the intellectual elite. While certain men of science such as Thomas Huxley found spiritualism unscientific, and the séance largely a charade perpetrated by cunning mediums, many others were willing to accept that the séance provided evidence for a world that existed beyond our everyday senses, a world that needed to be examined with scientific rigor.[57]

Myers, who had been educated at Cambridge and was a poet and a classicist, believed that the discoveries produced during the séance such as telepathic transmission must "revolutionize our whole attitude towards the question of an unseen world, and of our own past, present, or future existence therein."[58] As Leigh Penman has shown, Myers coined the term "supernormal" to describe just the kind of phenomena that could be witnessed during a séance. The term was important for Myers because it stressed that such a phenomenon did not contravene natural laws but, rather, "exhibits the action of laws higher, in a psychical aspect, than are discerned in action in everyday life." And for Myers, it was the science of evolution that could make sense of these supernormal phenomena. He argued that they were indications of a "more advanced stage of evolution."[59]

What Myers meant was that the progressive development implied in the evolutionary story of life provided a framework for thinking about how that life will continue after death. Central for Myers's understanding of that future life was that the most important process in the evolution of humans is that which is relevant to the development of thought and consciousness. In this regard, he argued, humans are still in a very early stage of development, given their failure to grasp the precise nature of the unseen spiritual world. He likened the current state of the human intellect with regard to its own development to that of a caterpillar sitting on a cabbage leaf trying to imagine how its own history relates to its future as a butterfly. Perhaps on witnessing a butterfly the caterpillar might be able to grasp the "imaginal characters," meaning the "points of structure which indicate that the larva has descended from an imago, or perfect insect, and is destined in his turn to become one himself."[60] But just as the caterpillar seems to grasp this future life, the butterfly flies away, leaving the caterpillar alone on the cabbage leaf, more confused than before. "This is exactly what I hold to have happened in the history of human evolution," claimed Myers, who posited that the history of human intellectual evolution includes such moments of sudden discovery, when it became possible to receive signs about humanity's evolutionary progress on a higher plane. He pointed to unexpected and swift advancements in geometry, mathematics, and music, which have occurred "without apparent heredit[ar]y cause" and therefore "indicate some access of energy outside the order of purely terrene evolution."[61]

Understandably, Myers was vague about what exactly the evolutionary future will hold after our terrene existence, given that we are like the caterpillar that is unable to examine clearly what will happen when we leave our current physical forms. But he stressed that we must simply assume that "spiritual evolution will follow the same laws as physical evolution" and that the evolutionary life of humanity will be continuous from its previous existence. In other words, there would be progress in the next life, but it would not come without an effort, without a struggle. Myers was clearly concerned that his readers might "sink into a self-absorption," in the trust that a future paradise will necessarily await them.[62] At the same time, he argued that there was little reason to worry about the inevitability of heat death: given that "man's soul [will] grow forever, it matters no more how many solar systems she wears out than how many coats."[63] Accepting the spiritual evolution of humanity meant that humans are therefore not limited by the earthly sphere, and that their destiny is "not planetary but cosmical, and . . . without an end."[64] This would also suggest that evolution is not partial or truncated but truly universal, extending backward in time to the origins of the cosmos and forward in an endless advance.[65]

It was certainly not much of a leap to take Myers's rather deistic view of the future life into a more explicitly Christian direction. There were various forms of theistic evolutionism in Victorian Britain, perhaps most notably promoted by the self-proclaimed Darwinian Charles Kingsley (1819–1875), who argued that the evolution of life was proof of God's absolute power, a view that had become so orthodox it was expressed by the future archbishop of Canterbury Frederick Temple (1821–1902) in his 1884 Bampton lectures.[66] For many Christians it was not much of a stretch to accept that evolution was a process that was begun by God in an initial act of creation. God may not have made all things, in Temple's words, but "He made them make themselves."[67] But fully explicating how such a view of evolution conformed to the Mosaic narrative of the Old Testament was no easy task. The Catholic convert St. George Jackson Mivart (1827–1900), for instance, argued that although humanity's physical form was produced entirely by evolutionary processes the soul was created in advance by God.[68] So there was a kind of dual origin for humans, according to Mivart's scheme: one origin for the soul, and another for the physical form. But what did such a view suggest about the processes of evolution after death? Was the soul not also subject to evolutionary processes even if created by God, indeed, because it was created by God?

These were the sorts of questions that the Unitarian Reverend John Page Hopps believed the study of supernormal phenomena helped to answer, because such questions suggested evolution was at work from the very beginning and continued on into the infinite future. In *The Future Life* (1884), Hopps argued that, when applied not just to the past but to the future as well, evolution confirmed the providential and progressive narrative he believed was central to Christian doctrine. The problem with the Darwinian history of humans as it was currently conceived, according to Hopps, was that it stipulated only one rather depressing story of life. That story was one of constant struggle leading to the formation of races and the creation of nations, which have occurred "with only groaning and pain." This was a problematic story, Hopps explained, because the great doctrine of "the survival of the fittest" that underpins it has really only been "half elaborated and half understood."[69] What was missing was a recognition that there is another meaning that must be ascribed to this doctrine, a meaning that is discerned from grasping the "ever recurring sigh of nature"—namely, the moment of death, which is also a moment of "redemption." Hopps argued that this moment of redemption foreshadows a new development, "a new departure in this marvelous process of evolution. It is as though the beautiful moth could be heard moaning and sighing through the shell of the chrysalis."[70]

Like Myers, Hopps saw in the transformation of the larval to the moth an analogy for a largely unseen but incredibly important stage in humanity's evolutionary history. He argued that we therefore have to carry the "glorious process [of evolution] onward into the unseen, in order to obtain one of the most rational and delightful conceptions of a future life. That life, properly understood, is only another step in the wonderful development of man's being: it is evolution still, but evolution into and in the sphere of the mind."[71] He explained that just as the "earth-process" has witnessed the creation of civilized humans out of a brutish origin, the further evolution of humans into the "spirit-world" may be just as dramatic a transformation. For Hopps, this notion was particularly reassuring because it raised "to an indefinite degree our conception as to what man really is, what he is living for, and what he may become." Accepting that human life did not end with death meant that "man" was actually a very different creature than was previously understood. According to Hopps "[he is] an altogether different kind of being, and the world is to him an altogether different kind of world. The whole out-look is changed. The whole calculation is altered."[72]

For Hopps, the reason the "whole out-look is changed" is because an evolutionary process that continues into the spiritual world means the theory is now consistent with Christianity, as it does away with the kind of dual origin that was necessitated by works like Mivart's. "One might say," Hopps argued, "that, for the first time in Christendom the human mind is coming to the possession of itself."[73] By grasping the spiritual dimension of evolutionary theory, it was now possible, according to Hopps, to understand the meaning behind the suffering and the pain. The possibility of a future life "harmonizes with all we know now of the Providence of God, and the laws of this house of the Lord in which we are today. It is a thought that throws light on most of the dark problems connected with the inequalities and sorrows and struggles of human existence here. It is a thought that makes all life, both here and there, an orderly and progressive development; the great transition called 'death' being only one step in the wonderful unfolding of human existence."[74]

It could be argued that the evolutionary writings of both Myers and Hopps represented a fairly extreme view of evolution, one that was therefore not taken seriously beyond a small audience of spiritually inclined naturalists. But it should be noted that, in making their case for the extension of the theory of evolution to the afterlife, Myers and Hopps appealed to the authority of one Alfred Russel Wallace. Indeed, it was not long after Wallace apparently posited an entirely natural origin for humanity at the ASL that he became interested in, and quite convinced of, the existence of spiritual phenomena, so much so that he realized that spiritualism helped explain some aspects of the evolution of life that were left unaccounted for by evolution by natural selection.

Wallace first went public with his views about spiritualism in *The Scientific Aspect of the Supernatural*, published in 1866. As well as putting forward an argument in favor of the evidence for, and therefore scientific validity of, spiritual phenomena, Wallace set out to establish continuity between the seen and unseen worlds. He believed that spiritualism enabled a truly holistic evolutionary cosmology and helped to explain certain perplexities that had arisen in his attempt to understand the evolution of life in its entirety. Now he argued that, if the law of the "survival of the fittest" explained the evolution of life for the organic world, it was the law of the "progression of the fittest" that explained the development of the human mind and its continued evolution beyond humanity's earthly existence.[75] Wallace stressed, however, that the spiritual dimension he was adopting did not negate his previous theory of evolution but, rather, completed it.

This last point was made explicitly in his widely read book *Darwinism* (1889), where Wallace proposed his spiritualist solution to the fundamental conundrums of human intellectual evolution that could not be explained by a strict adherence to natural selection.[76] After providing a fairly orthodox summary of how evolution has been shaped by natural selection, Wallace argued that there were limits to what natural selection could ultimately explain. In outlining those limits, he argued that it made sense to think of the history of all life as advancing through three stages, with each stage representing a particularly transcendent moment when a power beyond that of natural selection was imposed on the development of organic life, thereby ushering in immense transformations. The first stage was the moment when organic life originated as some sort of protoplasm emerging from inorganic material. This change had typically been explained by an increase in the complexity of chemical compounds. But Wallace now believed that such an explanation was unable to account for the origins of *living* protoplasm—that is, "protoplasm which has the power of growth and of reproduction, and of that continuous process of development which has resulted in the marvelous variety and complex organization of the whole vegetable kingdom." This was a sign for Wallace that there must be "a new power at work, which we may term *vitality*, since it gives to certain forms of matter all those characters and properties which constitute Life."[77]

The second stage "is still more marvelous," according to Wallace, because it involves the origins of consciousness, which distinguishes the animal and vegetable kingdoms. He claimed that it is "altogether preposterous to assume that at a certain stage of complexity of atomic constitution, and as a necessary result of that complexity alone, an *ego* should start into existence, a thing that *feels*, that is *conscious* of its own existence." There was quite simply no possible explanation relying on natural selection that could account for the arrival of a "being whose nascent consciousness has gone on increasing in power and definiteness till it has culminated in the higher animals." The animal that develops a consciousness able to think about its own existence simply cannot be explained with reference to a struggle for existence alone.[78]

The third stage in Wallace's evolutionary epic referred to the sudden existence of certain intellectual, cultural, and moral qualities in humans that separate them from "the brutes and open up possibilities of almost indefinite advancement."[79] Natural selection could not entirely account for the evolution of mind, because of the fact that certain mental faculties arose, such as mathematics and music, that did not exist at all in early societies and yet appeared quite "suddenly and in perfect development in the higher civilized races."[80] These faculties, moreover, were developed only by certain members of those societies, meaning that only a few were capable of pursuing these special faculties to their fullest expression. They could, therefore, be explained only by "something which we may best refer to as being of a spiritual essence or nature, capable of progressive development under favorable conditions."[81]

By accepting the intervention of a "spiritual essence of nature" to these moments of sudden complexity, Wallace believed that he was finally making sense of human evolution in connection with a larger history of life. He was therefore now able to come to terms with the special intellectual capabilities apparent in humanity's evolutionary history by merging a naturalist and spiritualist approach. This meant that the special characteristics unique to humans such as "the love of truth, the passion for justice"—along with the impressive capabilities of certain intellects to grasp the operations of nature—were not accidental products of a blind law of nature but, rather, signs of "the workings within us of a higher nature which has not been developed by means of the struggle for material existence."[82]

Wallace was adamant, however, that the facts he supplied in support of humanity's spiritual nature were "not in any way inconsistent with the theory of evolution" but were in fact "dependent on those fundamental laws and causes which furnish the very materials for evolution to work with." And, perhaps more important, Wallace suggested that this completely scientific view of humanity's evolutionary history would be a relief for many who thought that the grand evolution of life would simply come to an end with the burning out of the sun. Unlike those who still held on to the "soul-deadening belief" implied by an entirely materialist view of human evolution, Wallace argued that those accepting "the existence of a spiritual world" now finally had an evolutionary scheme that was worthy of the marvelous nature of life's history. It was possible to "look upon the universe as a grand consistent whole adapted in all its parts to the development of spiritual beings capable of indefinite life and perfectibility." Moreover, now the "whole purpose . . . of the world," from its complex physical structure on down to the slow and grand evolution of all its life, could be grasped as leading inevitably up to "the development of the human spirit in association with the human body."[83] Human evolution now had a meaning worthy of the process.

THE METAPHYSICS OF EVOLUTION

Coming to terms with a naturalized human origin was no easy task. Darwin did much to engender a revolution in thinking about the development of organic life, including that of humans, but his own theory of natural selection was largely rebuffed by general readers and by the large majority of naturalists. Fears of what "going the whole orang" meant for humanity in general put a premium on narratives of evolution that did what Darwin's largely did not; that is, provide a grand meaning for the evolutionary process that could help compensate for the loss of a supernatural origin. A range of evolutionary theorists and writers sought to do this, from supposedly devoted Darwinians to evolutionary spiritualists, from degenerationists to eugenicists. From the study of how their views of the evolutionary past informed the possible future direction of evolution, it is possible to see how they used the future as a space to project an ultimate purpose that underpinned the evolutionary process.

That purpose deviated depending on the given author, but they all seemed to agree that eventually humans would be able to overcome the Darwinian struggle for existence; that is, overcome those natural constraints that made humans a part of the natural world in the first place. However, there is no doubt that the possibility of degeneration and heat death undermined any unproblematic progressive narrative leading to an earthly utopia. And there was certainly an alternative school of evolutionary thought led by Huxley that argued that human society needed not to conform to the processes of evolution but, rather, to seek protection from their destructive forces.[84] Yet for a great many evolutionists, like those discussed throughout this chapter, taking over the processes of nature suggested the possibility of a transcendence of the limits imposed by the laws of science, whether of biology or of physics. For them, humanity's evolutionary future promised to occur on a higher plane of existence. This was a message that was explicit in Wallace's cosmic evolutionary theism, but it was also implied in Lankester's theory of degeneration and even in the evolutionary narrative that Galton used to explicate his program of eugenics.

There is no doubt that theories of evolution were informed by the concept of progress, as Michael Ruse has detailed in his survey of the development of evolutionary biology in *Monad to Man*.[85] Ruse highlights notions of social and political progress, but the focus of this chapter is on the centrality of metaphysical progress, the progress that is entailed in gaining a more fundamental understanding of the nature of humans. From this perspective, evolutionists were not simply replacing a supernatural myth of creation with a secular one that was more conducive to the political and social realities of modernity. They were more accurately contributing to the creation of evolution as a metaphysical and eschatological process that would eventually result in the merging of the human with the eternal cosmic mind.[86]

It is therefore important to recognize that the historicization of human origins in the second half of the nineteenth century was shaped not just by considerations of the past but by notions of the future as well. By focusing on the future projections that were established within the framework of evolutionary narratives of human origins, we can grasp more clearly the metaphysical vision of the universe underpinning the attempt to make evolution more palatable to a popular readership.[87] More generally, it makes it clear that, unfortunately for Darwin, managing hopes and fears was absolutely central to establishing an evolutionary theory of humans in the nineteenth century.

AFTERWORD

HISTORIOGRAPHICAL REFLECTIONS ON THE HISTORICIZATION OF HUMANS IN NINETEENTH-CENTURY BRITISH SCIENCES

THEODORE KODITSCHEK

In few respects is the past more of a foreign country than in its renderings of the past. Twenty-first-century conventional wisdom tells us that nineteenth-century Europeans were inveterate racists and imperialists, who became interested in the question of human origins mainly to acquire ammunition for exploiting and expropriating peoples of color—ranking them on a predetermined hierarchy of hereditary quality that placed such peoples at the bottom and themselves on the top. While not wrong, this assessment is superficial, since it fuels all too easily our own conceits of moral superiority as supposedly enlightened postcolonials. As historians have begun to look with greater dispassion into how human origins were understood in the nineteenth century, the stereotype has been greatly complicated, and a strange, alien environment has come into view. Epistemologically distant from our own mental universe, these nineteenth-century visions still point a broken path to our current ways of thinking. As we gaze upon this newly discovered intellectual landscape, we find ourselves emerging from our own Brixham cave of ignorance to survey an alien, yet also vaguely familiar, scene: a half-recognizable ancestral habitat filled with strange conceptual creatures that we can still identify as the distant progenitors of our own.

At the beginning of the nineteenth century, most educated men and women still subscribed to the six-thousand-year chronology of the Bible. Yet, by the end, the "deep time" of evolutionary thinking had become the consensus view. Even so, in the absence of population genetics, the entities envisioned as the objects of evolution were treated as types and races, rather than as the individuals, gene pools, and demes that evolutionary biologists quantify today. In the middle of this transition through "deep time" to evolution, a monumental edifice known as "Darwinism" arose, but as researchers began to examine humanity's actual influence, old answers began to crumble, and new questions were raised. Was Darwinian theory really the main paradigm for Victorian studies of evolution and human origins? Was it even central to prevailing notions of progress and race? Did "Social Darwinism" actually exist as a unifying framework for Eurocentric assertions of superiority, or was it merely an epithet polemically hurled at one another by late Victorian propagandists of diverse schools in the heat of debate?

Revisionist work on the history of nineteenth-century histories of human origins began in 1966, with John Burrow's *Evolution and Society*, a path-breaking study, where he argued that philology more than Darwinian biology was responsible for reopening the topic in an evolutionary analytical frame. This was followed by George Stocking's magisterial account of *Victorian Anthropology*. Stocking emphasized the ethnological traditions in which evolutionism was raised. Roger Bannister questioned the currency of "Social Darwinism" and raised doubts about whether it ever really constituted a coherent ideological force. Martin Rudwick explored the emergence of deep time; and the subsequent impact of the discovery of human antiquity was examined by Peter Bowler and Donald Grayson. On another track, Stephen Jay Gould and Nancy Stepan, to name only the most prominent, were exploring the role of race.[1] Important contributions came from the history of science, where Robert Richards traced the emergence of evolutionary psychology, Bernard Lightman interrogated popularizations of evolution, and James Secord demonstrated the important contribution to this endeavor that Robert Chambers initially introduced. Meanwhile, the Darwin industry continued to churn out its reams of pulp nonfiction, with important biographical syntheses by Adrian Desmond, James Moore, and Janet Browne, as well as many important specialized works.[2]

These pioneering contributions to scholarship might well be labeled "Historiography of Historicizing Human Origins 1.0." By the first decade of the present century, enough work had been done to offer a compelling first pass at revision, in which many of the old pieties had been dethroned. Race was a secondary concern for at least some of the investigators, who were genuinely interested in uncovering the "truth." Science and religion could no longer be viewed as inherently in conflict, since debates over evolution forged relationships between them that were a good deal more convoluted and complex. Disagreement ran rife between monogenists and polygenists, but absent a commonly agreed framework for discussion, it was not even clear how such controversies might be resolved. Evolution itself was a zone of contention, with no paradigm on which all evolutionists could agree. Charles Darwin (1809–1882) offered only one among several competing theoretical frameworks, and "natural selection" did not entirely dominate, even in his own books. Moreover, the search for human origins was not just a question of science, since it had an impact on many areas of nineteenth-century endeavor, popular culture, literature, and philosophy.

By the first decade of the present century, it seems fair to say that "Historiography of Historicizing Human Origins 1.0" had run its course, if only because the scholars who pioneered it were now beginning to pass from the scene. In the endnotes, pages, and chapters of this present volume a new generation of scholarship can be seen emerging, which we might label, "Historiography of Historicizing Human Origins 2.0." In many respects this historiography project is simply a continuation of its predecessor. The same themes, questions, and perspectives remain dominant, and the proliferation of scholarship lends a newfound depth and precision to the findings that are unearthed. In many ways, this is simply a natural product of the logic of subspecialization, which generates ever finer-grained pictures of an ever more closely observed scene. More rigorously trained, often under the mentorship of the first generation, these emerging young scholars are beginning to make the field their own. *Historicizing Humans* offers a sampling of some of their shorter contributions and, therefore, gives us a good opportunity to assess the current state of play in the field. Those who are looking for a deeper immersion need only follow the endnotes and bibliography, which offer a comprehensive listing of recent works, many by the authors of our chapters themselves.

The most obvious innovation of this newfound focus on detail is the increased attention that has been given to relatively understudied figures. In the past these men (this volume considers exclusively men) might have been dismissed as minor eccentrics, but here they are treated as exemplary figures, illustrating themes, patterns, and connections that have been hitherto overlooked. Thus Nanna Kaalund, in chapter 2, examines the life and work of John William Dawson (1820–1899), a Canadian paleontologist who combined Christianity and biblical scholarship in a unique and idiosyncratic way. In chapter 4, Maurizio Esposito and Abigail Nieves Delgado report on the South American explorations and promotional work of William Bollaert (1807–1876), a British collector and anthropologist, who combined a strong commitment to the economic development of the region with an utter contempt for the racial value of the indigenous inhabitants, whom he hoped to see replaced by European settlers as the process of development proceeded. Ian Hesketh, in chapter 8, sheds fascinating, illuminating light, on the redemptive visions of evolutionary transcendence purveyed by such nearly forgotten figures as W. Winwood Reade (1838–1875), Frederic Myers (1843–1901), and John Page Hopps (1834–1911), as well as such better known (but often mischaracterized) individuals as Alfred Russel Wallace (1823–1913), Francis Galton (1822–1911), and E. Ray Lankester (1847–1929).

This welcome attention to evolutionism's outliers does not preclude reexamination of its more canonic personalities. Thus, Gregory Radick, in chapter 6, enlivens these pages with a closely observed analysis of Darwin on emotions. By careful detective work in a welter of diverse sources, he persuasively shows how this seemingly arcane study was driven by the author's own emotional imperative to offer persuasive proof of the hereditary unity of humankind. Correspondingly, in chapter 3, Efram Sera-Shriar reconsiders Edward Burnett Tylor (1832–1917), not in retrospect as the founding father of "modern anthropology" but as a young man traveling in Mexico whose theory of "survivals" was worked up from the field observations that he made. This then opens the way to another significant innovation, as the volume registers a significant geographical shift from the discovery of human antiquity in Britain and Europe to the dispersed and disparate excavations and conversations that were conducted in Canada, Latin America, and the Indian subcontinent.

If there is a connecting thread that unites the majority of these variegated chapters, it is the need for adopting multiple perspectives. Close interrogations of specific themes and individuals enable us to see with greater clarity that the shift to "deep time" in understanding human origins did not enthrone any single theory or worldview. In a sensitive and intricate microanalysis, Chris Manias, in chapter 1, shows how the Victorians were forced to reevaluate the character of their distant ancestors once they realized that these protohumans had coexisted with the large and dangerous extinct animals, whose bones were scattered in near proximity at archaeological sites. What skills and courage must the Pleistocene European have possessed to subdue these dangerous monsters, and at what point did the human become the predator rather than the prey? What did recently discovered cave paintings reveal about human interaction with such animals—and about the earliest, most primitive versions of religion, aesthetics, and morality? Were humans responsible for the disappearance of large mammals, as their growing capabilities of technological mastery enabled them to domesticate a small minority of useful species, while attacking and killing off the rest?

Although we think of paleontology as a detheologizing endeavor, Nanna Kaalund shows that deep time did not necessitate abandonment of the mosaic chronology, since men like Dawson were able to interpret the biblical "days of creation" as metaphors for the epochs that geologists were discovering in the history of sedimentary rocks. In this manner, Genesis and geology could be cozily reconciled, and religion could be conveniently treated as complementary to science, in a manner that suited the needs of the embryonic institutions of colonial erudition. Under this banner, Dawson was able to reassert the case for monogenesis without ever subscribing to the potentially heretical premises of Darwinian biology. As with Dawson and the paleontologists, Tylor was also an uncompromising advocate of monogenesis, and he correspondingly read the archeological record of early humans as a story of steadily increasing technological and intellectual progress. Yet, so far from reconciling science with religion, his goal was to turn the study of religion itself into a scientific subject—even as he depended on missionary reports for evidence of the "comparative method" that enabled him to deploy modern savages as empirical proxies for early humankind. Here then, in a manner that differed greatly from Dawson, Tylor also found himself side-stepping rather than fully embracing the Darwinian paradigm, as the object of evolution was shifted from biology to culture.

For Bollaert in South America, the question of evolutionary progress was an open-and-shut case since he regarded the indigenous Americans as hereditarily inferior—an irreclaimable and hopelessly dying race. To buttress these prejudices, he invoked the polygenetic premise, yet he was quite prepared to combine this with Darwinian mechanisms of natural selection, in hopes that the advent of European imperialism would repopulate the American continent and drive the indigenous Americans to their well-deserved extinction. Esposito and Nieves Delgado show that Bollaert, in further contrast both to Tylor and to Dawson, was much less driven by either science or religion than he was by an avaricious interest in capitalist development and a desire to demonstrate that evolution and imperial aggrandizement were aligned.

Bollaert's crassly instrumental applications of evolutionary theory to justify the exploitation and expropriation of indigenous Americans reminds us that the discourse of evolution was never entirely about the past. His conviction that capitalist development was the inevitable way of the future deeply colored his reading of the story of human origins, which became but a prologue to the triumph of men such as himself. In Ian Hesketh's account of the five late nineteenth-century utopian visionaries we see loftier exemplifications of a comparable move. Where Bollaert, the entrepreneurial collector, sought to maximize investor profits, Hesketh's intellectuals considerably upped the ideological ante, with their towering redemptive prophecies. In their accounts, the evolutionary process was neither random nor meaningless, as the "survival of the fittest" became tantamount to the ascendancy of the good. Not, of course, that this path would be easy, since the violence of crude struggle would have to be transcended, and the retrogressive threat of degeneration always needed to be overcome. Still, surmounting these difficulties only sweetened the ultimate triumph by which progress would eventually win out in the end. Once again, in these accounts, religion and science were inextricably mingled, although it was a uniquely secularized version of religion that was advanced this time.

With all these intricate cross-cuttings of theory, method, discipline, moralizing, motive, and result, it is difficult to draw any conclusions for this volume as a whole. Indeed, the only conclusion might be that there can be no single conclusion. For a field that was in such rapid transition, in which hypotheses were being radically revised and again reconsidered, and in which geologists had to confront the Bible, biologists were waylaid by language, ethnologists took to measuring skulls, paleontologists became entertainers, and utopian dreamers lofted gossamer visions on butterfly wings. Everything here can seem so intricate, idiosyncratic, and fluid as to defy any hope that generalizations can be made. Nevertheless, when one stands surrounded by such a welter of diverse trees, it is important to remember that one is in a forest, and every forest (if one approaches it at sufficient aerial height) will eventually reveal its ecology.

In chapter 7, Helen Kingstone offers one potential catapult that might raise us up to the requisite conceptual altitude. If the discovery of "deep time" was the single connective thread that unites all the other chapters, then it might repay the effort to consider how this discovery might have altered attitudes towards the shallow time of conventional historical narrative. How far did the comparative method provide a bridge between these distances, enabling ethnological (or anthropometric) study of modern "savages" to serve as a window into deep antiquity, or vice versa? Connected to this theme in Kingstone's chapter is the question of stadial development—a paradigm originally devised by the sociohistorians of the Scottish Enlightenment (Adam Smith [1723–1790], John Millar [1735–1801], and Adam Ferguson [1723–1816]), but which is said to have atrophied in the nineteenth century under the glacial weight of sedimentary gradualism that had been introduced into evolutionary thinking by the analytic model of geology. To what extent, Kingstone asks, did some version of stadialism survive amid the sedimentary discourse of the nineteenth century? Kingstone is wise to begin with Walter Scott (1771–1832), who, as she notes, was steeped in the stadial theories of his Scottish predecessors, but who transposed them onto a different *narrative* fabric in his historical novels. Scott accomplished this remarkable feat by cannily juxtaposing different regional timescales of modernization that separated the histories of the Highlands and Lowlands in Scotland and those of England and Scotland on the larger Isle. Encountering their divergent development at differential velocities, Scott found in the resultant clash a spectacle-raising drama that could fuel his story plots of conflict, and reconciliation. It is perhaps unsurprising that I find

Kingstone's arguments here to be entirely convincing, since they recapitulate some that I published nearly a decade ago, often using the same examples that appear in her work.[3]

Nevertheless, I am less persuaded by Kingstone's forays into the nineteenth century with her subsequent analyses of Thomas Carlyle (1795–1881) and Galton. It is not that Kingstone is wrong in her renderings of these figures but, rather, that their work does not sufficiently engage with the problem of stadialism, to tell us much about how it might have been used.[4] A far better example, it seems to me, can be found in the work of Karl Marx (1818–1883) and Friedrich Engels (1820–1895), who quite explicitly posed themselves the challenge of updating the stadialism of the Scottish sociohistorians by recasting it in a "progressive" evolutionary frame. What eighteenth-century stadialism lacked was a mechanism to explain the *transition* from one stage to the next. It is this that Marx and Engels supplied, with their dynamic dialectic between "material forces and social relations of production," "contradictions within the mode of production," and "revolutionary transformation from one mode of production to the next." In this manner, Marx and Engels were able to endow their stages with an evolutionary thrust of progressive directionality—albeit at the cost of conjuring up some of the same teleological overtones that Hesketh detects in his more conventional evolutionary epics.[5]

The difference between the evolutionism of Galton or Wallace and the Marx-Engels team is that the former envisioned gradualist workings of biosocial accretion, and the latter summoned up a purely *social* transformative dynamic that relied on the exercise of human agency. This was a very different solution to the problem of reconciling longue durée evolutionary change with "shallow time" narrative action. It is a measure of the insularity of the British evolutionists that not even the single socialist among them (Wallace) felt compelled to engage with the challenge of Marx and Engels's work.[6] Nevertheless, it would be a mistake to conclude that the British evolutionists had no recourse for dealing with this conundrum, even if they never explicitly formulated it in a theoretically rigorous way.

On the contrary, I would argue, it was the very vagueness and uncertainty of their theoretical apparatus that enabled the mainstream evolutionists to finesse this problem, which they could not resolve. Here we run up against the dilemma of "Darwin," and the need to reassess the contemporary influence of his work. We must ask why no amount of revisionist hedging and qualification will ever entirely displace him from the story of evolution's ascendancy? The reason, I believe, is that the sheer rigor, simplicity, and elegance of natural selection ensured that, once formulated, the theory would never go away. Radick's brilliant detective work in this volume clearly demonstrates that when it came to conceptual clarity and intellectual integrity, Darwin was simply in a class by himself. He would not compromise his scientific judgment even when motivated by emotional and ideological commitments that he desperately hoped to align with his objective research. But Darwin was profoundly moved by emotional and ideological considerations, such that the amorality of natural selection troubled even him. As critics identified more of the problems in the theory, he allowed a strain of neo-Lamarckism to creep back into his later writings, blunting the force of random variation and natural selection with a teleologically inflected expectation that the evolutionary trajectory would prove to be morally progressive in the end.[7]

Where Darwin capitulated reluctantly to these moralizing impulses, many of his fellow practitioners were eager to go the whole hog, lacing the sober logic of natural selection with teleological trip wires, whether Lamarckian, orthogenic, recapitulationist, or openly deistic, that endowed the evolutionary process with an ethically directional thrust. This is the story that Hesketh tells in this volume, but the pattern was not limited to his visionary epicists. It was much more widespread, as Michael Ruse and others have shown.[8] Indeed, in the decades before August Weismann's demonstration of the segregation of the germplasm (and absent awareness of Gregor Mendel's experiments with peas), the dearth of any coherent theory of heredity meant it was open season for speculation, and some residual belief in the inheritance of acquired characteristics became almost the default mode for many practitioners. Darwin provided one example with his theory of pangenesis, but others such as George Romanes (1848–1894), Samuel Butler (1833–1902), and the Duke of Argyll, George Douglas Campbell (1823–1900), were prepared to go much further.[9]

The classic case for this mixing and matching of evolutionary mechanisms was that of Herbert Spencer (1820–1903), and there is good reason to think that Spencer's somewhat loose and opportunistic amalgam of Lamarck with Darwin was, given Spencer's enormous popularity and influence, the most widely accepted evolutionary heuristic during the second half of the nineteenth century.[10] Among the figures discussed in this volume, Reade very likely falls in this category, as (with qualifications) does Wallace. Lankester repudiated this amalgam only after he came under the influence of Weismann, and even Tylor can be read as harboring some sympathies. If we allow God to make acquired characteristics heritable, then Dawson, George Jackson Mivart (1827–1900), and Hopps will join this crowd.

The point of this exercise is to demonstrate that nearly all of these idiosyncratic versions of evolution shared a common feature, in that the need to accommodate directionality and progress necessitated the amalgamation of mechanisms and the multiplication of timescales. When inheritance, God, or some mysterious orthogenic principle sped up the pace of human biosocial development, the velocity of evolution correspondingly quickened. Progress would then proceed more rapidly, although degeneration might also set in at an accelerated tempo. As a result, it is somewhat anachronistic to juxtapose our modern dichotomy between the "deep time" of evolution and the "shallow time" of narrative conjuncture, when we analyze nineteenth-century evolutionary works. For many of these Victorians, much of human evolution marched neither to the slow sedimentary rhythm of natural selection nor to the fast-paced purposive action of modern history. On the contrary, it was conceived in composite fashion as a multispeed, biosocial process, in which individuals and races were both the products of heredity but also were able to alter their individual or racial character by limited acts of purposive agency.[11]

For Victorian liberals, which included many who voted Conservative, no value was more sacred than the idea of "improvement," and any evolutionary theory that hoped to gain widespread acceptance would have to accommodate itself to this aspiration. To be sure, the capacity for self-help was widely regarded as a racial, and also gendered, trait, well developed in white Anglo-Saxons (especially of the middle class) and scarcely evident in savages and peoples of color, who were regarded as backward in their inherent evolutionary potentiality. But some capacity for improvement could scarcely be denied to such people, or at least to their descendants in a distant future, when the gradual workings of use-inheritance (or divine providence) might bear fruit in cumulative cultural/hereditary advance. This point, I would argue, has considerable importance, since it helps to explain why the not exclusively Darwinian evolutionism that was popular during the Victorian period played a critical role in legitimating the project of liberal imperialism. For this was an age when indigenous peoples could no longer be openly enslaved or exterminated, and even legal discrimination was difficult to justify. At the same time, British imperialists coveted the land, and peripheral capitalists needed to exploit the hard labor of such groups. In an age of democratic advancement, moreover, it was necessary to find some pretext for denying them the status of a fully fledged citizenry. By justifying, in the name of science, exclusions and discriminations that could no longer be entirely exacted by law, evolution played a critical role in reconciling liberal Britons to the empire, while turning the empire into a polity that could be ruled in authoritarian fashion, even as it appeared to be governed in a formally liberal way.[12]

For imperial administrators, evolution was a gift that helped to clarify their civilizing missions, while providing them with the conceptual tools to formulate graduated strategies of coercive domination. In this regard, Bollaert, the overt exterminationist, was something of a throwback, and Herbert Hope Risley (1851–1911), whom Thomas Simpson quotes in his epigraph in chapter 5, exemplified the way of the future. Risley's mission was to organize a vast census of the entire Indian population, dividing it "scientifically" into its constituent races, so that each could be assigned its proper imperial function, as research certified its proper place on the evolutionary scale. Risley's work was empirically informed by a massive collection of anthropometric data, but as Simpson shows, it was also grounded in several generations of philological study that went back to the original "Aryan" hypothesis of Sir William Jones (1746–1794). As British power spread more widely throughout the Indian subcontinent, subsequent linguistic scholarship in Madras and Bombay purported to uncover a deeper, more primitive pre-Aryan layer that was still evident in the low caste, and tribal populations of the southern and western interior.[13] Here, as elsewhere, culture and biology were intermingled, as racial character was seen as exemplified in language, and language was taken as a proximate index of race. As the Raj became less dependent on (and less interested in) the collaborative support of the Bengali Bhadralok, an army of ethnologists awoke to the potential labor (and military service) that could be extracted from the supposedly more primitive (and more compliant) lower tribes, and castes.[14]

Perhaps the most interesting part of Simpson's chapter is his discussion of Indian responses to Darwin, which often precipitated an effort to fuse him with formulations from the Hindu tradition, or even claims that Hindu sages had anticipated Darwin's work. Yet, I would suggest taking this argument one step beyond Simpson, since Hindu evolutionism was a reaction to the power of officials such as Risley, as much as it was a response to the ideas of Darwin.[15] In the hands of a few of the more extreme protonationalists, it even led to claims that the Indians were the superior race. By the end of the century, it was becoming clear that racial evolutionism was a game that not only imperialists could play. Indeed, one might go so far as to trace the roots of some of today's toxic postcolonial Hindu and Muslim ethnocentrism to this extra-European appropriation of the racial-evolutionary discourse that the British imperialists had first introduced.[16]

So, in the end, we can support the conclusion that the quest of British evolutionists to understand human origins was inextricably interconnected with empire and race. It is essential, however, to recognize that the relationship was complicated, and that evolutionism was more than simply an ideological alibi for discrimination and imperial rule. Indeed, ideas about race and evolution did not merely justify imperial dominance but actually helped to construct that dominance, for they shaped the form that imperialism took. Conversely, imperialism, and the need for capitalist exploitation on the periphery, was an important factor in ensuring that evolutionary approaches to human antiquity would rise to the forefront of Victorian intellectual discourse, would assume center stage in the national consciousness, and would body forth in the particular theoretical dress as reported in our chapters. As we move with greater certainty into understanding this relationship between ideas and their sociopolitical consequences, the landscape of nineteenth-century human origin studies will begin to look less alien and strange. We are fortunate that the present volume makes an important contribution toward moving us forward in pursuit of this goal.

NOTES

INTRODUCTION

1. Alfred Henry Huth, *The Life and Writings of Henry Thomas Buckle*, 2 vols. (London: Sampson, Low, Marston, Searle & Rivington, 1880), 1:255–56.
2. Henry Thomas Buckle, *History of Civilisation in England*, 2. vols. (London: J. W. Parker, 1857). See also Ian Hesketh, *The Science of History in Victorian Britain* (London: Pickering and Chatto, 2011), 13–14.
3. The term "deep time" was not used by Victorians, and it was first used in the historiography by John McPhee in his important work *Basin and Range* (New York: Farrar, Straus, Giroux, 1981), 20. See also Kirsty Douglas, *Pictures of Time Beneath: Science, Heritage and the Uses of the Deep Past* (Clayton: Csiro Publishing, 2010), 10–13; Amy Ione, *Art and the Brain: Plasticity, Embodiment and the Unclosed Circle* (Leiden: Brill, 2016), 18–20.
4. Peter Bowler, *The Invention of Progress: The Victorians and the Past* (Oxford: Basil Blackwell, 1989), 1.
5. George Grote, *A History of Greece: From the Earliest Period to the Close Generation Contemporary with Alexander the Great*, 12 vols. (London: John Murray, 1846–1856), 1: 249, 424, 428.

6. For more on the evolution of the term "historicism," see Mark Bevir, "Introduction," in *Historicism and the Human Sciences in Victorian Britain*, ed. Mark Bevir (Cambridge: Cambridge University Press, 2017), 1–20.
7. For a more nuanced understanding of the boundaries between science and religion, see Frank Turner, "The Victorian Conflict between Science and Religion: A Professional Dimension," *Isis* 69 (1978): 356–76; John Brooke and Geoffrey Cantor, "Whose Science? Whose Religion?" in *Reconstructing Nature: The Engagement of Science and Religion*, ed. John Brooke and Geoffrey Cantor (Oxford: Oxford University Press, 1998), 43–72; Bernard Lightman, "Victorian Sciences and Religion: Discordant Harmonies," *Osiris* 16 (2001): 343–66; Peter Harrison, "'Science' and 'Religion': Constructing the Boundaries," *Journal of Religion* 86 (2006): 81–106.
8. Bevir, "Introduction," 2–3.
9. Martin Rudwick, *Bursting the Limits of Time: The Reconstruction of Geohistory in the Age of Revolution* (Chicago: University of Chicago Press, 2005), 276.
10. McPhee, *Basin and Range*, 20.
11. Martin Rudwick coined the term "deep history" in his book *Earth's Deep History: How It Was Discovered and Why It Matters* (Chicago: University of Chicago Press, 2014).
12. Efram Sera-Shriar, *The Making of British Anthropology, 1813–1871* (London: Pickering and Chatto, 2013), 90–94.
13. Robert Knox, *The Races of Man: A Fragment* (London: Henry Renshaw, 1850), 9.
14. The most comprehensive works on the link between secularism and scientific naturalism are Frank Turner, *Between Science and Religion: The Reaction to Scientific Naturalism in Late Victorian England* (New Haven: Yale University Press, 1974); and Bernard Lightman, *The Origins of Agnosticism: Victorian Unbelief and the Limits of Knowledge* (Baltimore: Johns Hopkins University Press, 1987).
15. See Auguste Comte, *Cours de Philosophie Positive*, 6 vols. (Paris: Bachelier, 1830–1842).
16. For more on Comte's influence on developmental theories in nineteenth-century British sciences, see Hesketh, *The Science of History*, 17–18.
17. A. B. Van Riper, *Men among the Mammoths: Victorian Science and the Discovery of Human Prehistory* (Chicago: University of Chicago Press, 1993), 2–6, 247; James Secord, *Victorian Sensation: The Extraordinary Publication, Reception, and Secret Authorship of* Vestiges of the Natural History of Creation (Chicago: University

of Chicago Press, 2000); Peter Kjaergaard, "Hurrah for the Missing Link! A History of Apes, Ancestors and a Crucial Piece of Evidence," *Notes and Records the Royal Society Journal of the History of Science* 65 (2011): 83–98. See also James Cowles Prichard, *Researches into the Physical History of Man* (London: John and Arthur Arch, 1813); Robert Chambers, *The Vestiges of the Natural History of Creation* (London: John Churchill, 1844); Charles Lyell, *The Geological Evidences of the Antiquity of Man* (London: John Murray, 1863); Thomas Huxley, *Evidence as to Man's Place in Nature* (London: Williams and Norgate, 1863); John Lubbock, *Pre-Historic Times: As Illustrated by Ancient Remains, and the Manners and Customs of Modern Savages* (London: Williams and Norgate, 1865); Edward Burnett Tylor, *Primitive Culture: Researches into the Development of Mythology, Philosophy, Religion, Language, Art and Custom,* 2 vols. (London: John Murray, 1871); and Charles Darwin, *The Descent of Man, and Selection in Relation to Sex*, 2 vols. (London: John Murray, 1871).

18. There is a lot of scholarship on the significance of voyages of exploration in shaping scientific understandings of the world. For examples, see George Stocking, *Victorian Anthropology* (New York: Free Press, 1987), 78–102; Janet Browne, "A Science of Empire: British Biogeography before Darwin," *Revue d'Histoire des Sciences* 45 (1992): 453–75; Janet Browne, *Charles Darwin: Voyaging* (London: Pimlico, 1995); David Livingstone, *Adam's Ancestors: Race, Religion, and the Politics of Human Origins* (Baltimore: Johns Hopkins University Press, 2008), 16–25; Adrian Desmond and James Moore, *Darwin's Sacred Cause: Race, Slavery and the Quest for Human Origins* (London: Penguin Books, 2009), 68–110; Sera-Shriar, *The Making of British Anthropology*, 43–51; Efram Sera-Shriar, "Tales from Patagonia: Phillip Parker King and Early Ethnographic Observation in British Ethnology, 1826–1830," *Studies in Travel Writing* 19 (2015): 204–23.

19. Mary Louise Pratt, *Imperial Eyes: Travel Writing and Transculturation* (Abingdon: Routledge, 1992), 5–8.

20. For more on the impact of alternative chronologies of human history, see Livingstone, *Adam's Ancestors*, 8–11.

21. Paul Pettit and Mark White, "Cave Men: Stone Tools, Victorian Science, and the 'Primitive Mind' of Deep Time," *Notes and Records: The Royal Society Journal of the History of Science* 65 (2010): 31; Kjaergaard, "Hurrah for the Missing Link!" 84.

22. Peter Bowler, *The Eclipse of Darwinism: Anti-Darwinian Evolution Theories in the Decades around 1900* (Baltimore: Johns Hopkins University Press, 1983); Peter

Bowler, *The Non-Darwinian Revolution: Reinterpreting a Historical Myth* (Baltimore: Johns Hopkins University Press, 1988); Bowler, *The Invention of Progress*; James Elwick, *Styles of Reasoning in the British Life Sciences: Shared Assumptions* (London: Pickering and Chatto, 2007); Robert Kenny, "From the Curse of Ham to the Curse of Nature: The Influence of Natural Selection on the Debate on Human Unity before the Publication of *The Descent of Man*," *British Journal for the History of Science* 40 (2007): 363–88; Gregory Radick, "Darwin's Puzzling Expression," *Comptes Rendus Biologies* 333 (2010): 181–87. See also Efram Sera-Shriar, "Human History and Deep Time in Nineteenth-Century British Sciences: An Introduction," *Studies in History and Philosophy of Biological and Biomedical Sciences* 51 (2015): 19–22.

23. Stephen Jay Gould, *Time's Arrow, Time's Cycle: Myth and Metaphor in the Discovery of Geological Time* (Cambridge, MA: Harvard University Press, 1987); David Oldroyd, *Thinking about the Earth: A History of Ideas in Geology* (Cambridge, MA: Harvard University Press, 1996); Ralph O'Connor, *The Earth on Show: Fossils and the Poetics of Popular Science, 1802–1856* (Chicago: University of Chicago Press, 2007); Rudwick, *Bursting the Limits of Time*.

24. For literature on nineteenth-century anthropology, see Stocking, *Victorian Anthropology*; Henrika Kuklick, *The Savage Within: The Social History of British Anthropology, 1885–1945* (Cambridge: Cambridge University Press, 1991); Henrika Kuklick, "The Theory of Evolution and Cultural Anthropology," in *The Theory of Evolution and Its Impact*, ed. Aldo Fasolo (New York: Springer-Verlag, 2012), 83–102; Chris Manias, *Race, Science, and the Nation: Reconstructing the Ancient Past in Britain, France and Germany* (Abingdon: Routledge, 2013); Douglas Lorimer, *Science, Race Relations and Resistance: Britain, 1870–1914* (Manchester: Manchester University Press, 2013). For literature on Britain and its empire, see Catherine Hall, *Civilizing Subjects: Metropole and Colony in the English Imagination, 1830–1867* (Chicago: University of Chicago Press, 2002); Andrew Thompson, *The Empire Strikes Back? The Impact of Imperialism on Britain from the Mid-Nineteenth Century* (London: Pearson Longman, 2005); Sujit Sivasundaram, *Nature and the Godly Empire: Science and Evangelical Mission in the Pacific, 1795–1850* (Cambridge: Cambridge University Press, 2005); Daniel Headrick, *Power over Peoples: Technology, Environments, and Western Imperialism, 1400–Present* (Princeton, NJ: Princeton University Press, 2010).

25. Sera-Shriar, "Tales from Patagonia," 209–10.

26. For more on the economic significance of India for nineteenth-century British imperialism, see Thompson, *The Empire Strikes Back,* 160–66.
27. Catherine Hall and Sonya Rose, "Introduction: Being at Home with the Empire," in *At Home with the Empire: Metropolitan Culture and the Imperial World,* ed. Catherine Hall and Sonya Rose (Cambridge: Cambridge University Press, 2006), 1–31. See also Hall, *Civilizing Subjects.*
28. Thompson, *The Empire Strikes Back,* 1–8.
29. James Secord, "Knowledge in Transit," *Isis* 95 (2004): 654–72.
30. There is a large body of scholarship on the relationship between science and the British periodical press. For example, see Jonathan Topham, "Scientific Publishing and the Reading of Science in Nineteenth-Century Britain: A Historiographical Survey and Guide to the Sources," *Studies in the History and Philosophy of Science* 31 (2000): 559–612; Geoffrey, Cantor et al., eds., *Science in the Nineteenth-Century Periodical: Reading the Magazine of Nature* (Cambridge: Cambridge University Press, 2004); Geoffrey Cantor and Sally Shuttleworth, eds., *Science Serialized: Representations of the Sciences in Nineteenth-Century Periodicals* (Cambridge, MA: MIT Press, 2004); Aileen Fyfe and Bernard Lightman, eds., *Science in the Marketplace: Nineteenth-Century Sites and Experiences* (Chicago: University of Chicago Press, 2007); Bernard Lightman, *Victorian Popularizers of Science: Designing Nature for New Audiences* (Chicago: University of Chicago Press, 2007).
31. The most comprehensive study of non-Darwinian forms of evolution is Bowler, *The Non-Darwinian Revolution*; Peter Bowler, *Darwin Deleted: Imagining a World without Darwin* (Chicago: University of Chicago Press, 2013). See also Sera-Shriar, "Human History and Deep Time," 19–22; and Ian Hesketh, "A Good Darwinian? Winwood Reade and the Making of a Late Victorian Evolutionary Epic," *Studies in History and Philosophy of Biological and Biomedical Sciences* 51 (2015): 44–52.

CHAPTER 1: CONTEMPORARIES OF THE CAVE BEAR AND THE WOOLLY RHINOCEROS

The research for this chapter was supported by a British Academy Mid-Career Fellowship (R117065: The Lost Beasts: International Palaeontology and the Evolution of the Mammals).

1. William Boyd Dawkins, "The British Lion," *Popular Science* (November 1882): 73–74.

2. John W. Burrow, *Evolution and Society: A Study in Victorian Social Theory* (Cambridge: Cambridge University Press, 1968); Thomas Trautmann, *Aryans and British India (*Berkeley: University of California Press, 1997).
3. See especially John McNabb, *Dissent with Modification: Human Origins, Paleolithic Archaeology and Evolutionary Anthropology in Britain, 1859–1901* (Oxford: Archaeopress, 2012); A. B. Van Riper, *Men among the Mammoths: Victorian Science and the Discovery of Human Prehistory* (Chicago: University of Chicago Press, 1993); Donald Grayson, *The Establishment of Human Antiquity* (New York: Academic Press, 1983).
4. As a note on terminology, this paper primarily focuses on understandings of the Pleistocene, the geological period just before the current era. This period included the Ice Ages and the Paleolithic, or "Old Stone Age"—a period in which humans were judged to have lived in small nomadic hunting groups, with only chipped stone artifacts.
5. Hélène Lafont-Couturier and Katia Bosch, eds., *Vénus et Caïn: Figures de la Préhistoire, 1830–1930* (Bordeaux: Réunion des musées nationaux, 2003); Michael Barany, "Savage Numbers and the Evolution of Civilization in Victorian Prehistory," *British Journal for the History of Science* 47, no. 2 (2014): 239–55; Theodore Koditschek, "Narrative Time and Racial/Evolutionary Time in Nineteenth-Century British Liberal Imperial History," in *Race, Nation and Empire: Making Histories, 1750 to the Present*, ed. Catherine Hall and Keith McClelland (Manchester: Manchester University Press, 2010), 36–55.
6. Particularly good studies on this are Martin Rudwick, *Worlds before Adam: The Reconstruction of Geohistory in the Age of Reform* (Chicago: University of Chicago Press, 2008), and Martin Rudwick, *Bursting the Limits of Time: The Reconstruction of Geohistory in the Age of Revolution* (Chicago: University of Chicago Press, 2005). See also Ralph O'Connor, *The Earth on Show: Fossils and the Poetics of Popular Science, 1802–1856* (Chicago: University of Chicago Press, 2007).
7. Discussed in Van Riper, *Men among the Mammoths*; and Grayson, *The Establishment of Human Antiquity*.
8. Irina Podgorny, "Fossil Dealers: The Practices of Comparative Anatomy and British Diplomacy in Latin America, 1820–1840," *British Journal for the History of Science* 46, no. 4 (2013): 647–74; Juan Pimentel, "Across Nations and Ages: The Creole Collector and the Many Lives of the Megatherium," in *The Brokered*

World: Go-Betweens and Global Intelligence, 1770–1820, ed. Simon Schaffer et al. (Sagamore Beach, MA: Science History Publications, 2009), 321–54.

9. Kirsty Douglas, *Pictures of Time Beneath: Science, Heritage and the Uses of the Deep Past* (Clayton: Csiro Publishing, 2010).
10. O'Connor, *Earth on Show*, 159.
11. Richard Owen, *A History of British Fossil Mammals and Birds* (London: John van Voorst, 1846), 182–83.
12. Richard Owen, *Description of the Skeleton of an Extinct Gigantic Sloth,* Mylodon Robustus (London: John van Voorst, 1842), 157–58.
13. William Buckland, *Reliquiae Diluvianae, or, Observations on the Organic Remains Contained in Caves, Fissures and Diluvial Gravel* (London: John Murray, 1824); discussed in Marianne Sommer, *Bones and Ochre: The Curious Afterlife of the Red Lady of Paviland* (Cambridge, MA: Harvard University Press, 2007).
14. Philippe Charles Schmerling, *Recherches Sur Les Ossemens Fossiles Découverts Dans Les Cavernes de La Province de Liège* (Liège: P. J. Collardin, 1833).
15. This process is discussed in Grayson, *The Establishment of Human Antiquity*; and Van Riper, *Men among the Mammoths.*
16. Charles Lyell, *The Geological Evidences of the Antiquity of Man* (London: John Murray, 1863), 1.
17. Lyell, *The Geological Evidences*, 264.
18. Thomas Jones, ed., *Reliquiae Aquitanicae: Being Contributions to the Archaeology and Palaeontology of Périgord and the Adjoining Provinces of Southern France* (London: Williams & Norgate, 1875).
19. Jones, *Reliquiae Aquitanicae*, 21.
20. Édouard Lartet, "Nouvelles Recherches sur la Coexistence de l'Homme avec des Grands Mammifères Fossiles Réputés Caractéristiques de la Dernière Période Géologique," *Annales des Sciences Naturelles, Quatrième Série: Zoologie* 15 (1861): 176–253.
21. Louis Figuier, *L'homme primitif* (Paris: Librairie de L. Hachette, 1870); Louis Figuier, *Primitive Man* (London: Chapman and Hall, 1870).
22. Peter Rowley-Conwy, *From Genesis to Prehistory: The Archaeological Three Age System and Its Contested Reception in Denmark, Britain, and Ireland* (Oxford: Oxford University Press, 2007).
23. William Boyd Dawkins, *Cave Hunting: Researches on the Evidence of Caves Respecting the Early Inhabitants of Europe* (London: Macmillan, 1874), 352.

24. For the concept of the "economy of nature," see Donald Worster, *Nature's Economy: A History of Ecological Ideas* (Cambridge: Cambridge University Press, 1994); John Glendening, *The Evolutionary Imagination in Late-Victorian Novels: An Entangled Bank* (Aldershot: Ashgate, 2016), 12; Margaret Schabas, *The Natural Origins of Economics* (Chicago: University of Chicago Press, 2005).
25. This has been particularly discussed in the case of the materialist prehistorian Gabriel de Mortillet—see, for example, Nathalie Richard, "Archeology, Biology, Anthropology: Human Evolution according to Gabriel de Mortillet and John Lubbock," *History and Philosophy of the Life Sciences* 34, nos. 1–2 (2012): 9–31.
26. Figuier, *Primitive Man*, 41.
27. Figuier, *Primitive Man*, 44, 52, 89.
28. John Lubbock, *Pre-Historic Times: As Illustrated by Ancient Remains, and the Manners and Customs of Modern Savages* (London: Williams and Norgate, 1865), 340.
29. Lubbock, *Pre-Historic Times*, 440.
30. Tim Murray, "Illustrating 'Savagery': Sir John Lubbock and Ernest Griset," *Antiquity* 83 (2008): 488–99.
31. Worthington George Smith, *Man, the Primeval Savage: His Haunts and Relics from the Hill-Tops of Bedfordshire to Blackwall* (London: E. Stanford, 1894), 56.
32. Smith, *Man, the Primeval Savage*, 47–48.
33. Smith, *Man, the Primeval Savage*, 59.
34. For a wider discussion of Dawson and his significance, see Nanna Katrine Lüders Kaalund, "Of Rocks and 'Men': The Cosmogony of John William Dawson" (this volume).
35. John William Dawson, *The Story of the Earth and Man* (New York: Harper and Brothers, 1873), 366–67.
36. Dawson, *The Story of the Earth and Man*, 368.
37. Henry Knipe, *Evolution in the Past* (London: Herbert and Daniel, 1912), 188.
38. Knipe, *Evolution in the Past*, 195, 196.
39. Oscar Moro Abadía, "Thinking about the Concept of Archive: Reflections on the Historiography of Altamira," *Complutum* 24, no. 2 (2013): 145–52; Eduardo Palacio-Pérez, "Salomon Reinach and the Religious Interpretation of Palaeolithic Art," *Antiquity* 84 (2010): 853–63.
40. H. G. Spearing, *The Childhood of Art* (New York: G. P. Putnam's Sons, 1912): 101.
41. William Boyd Dawkins, *Early Man in Britain and His Place in the Tertiary Period* (London: Macmillan, 1880), 345–47.

42. Richard Owen, *Palaeontology, or, a Systematic Summary of Extinct Animals and Their Geological Relations* (Edinburgh: Adam and Charles Black, 1860), 398.
43. Owen, *Palaeontology*, 401.
44. Lubbock, *Pre-Historic Times*, 328–29.
45. Lyell, *The Geological Evidences*, 23.
46. Figuier, *Primitive Man*, 171.
47. Figuier, *Primitive Man*, 344–45.
48. Knipe, *Evolution in the Past*, 214–15.

CHAPTER 2: OF ROCKS AND "MEN"

I am thankful to Peter Harrison and Ian Hesketh, who provided me with funds to travel to the Evolution and Historical Explanation: Contingency, Convergence, and Teleology (CHED-IRC) Conference in 2014, where I presented this material in its early stages. My thanks also extend to Bernie Lightman, Colin Coates, Ernst Hamm, and Jon Topham for their helpful advice and suggestions. My thanks also extend to Jon Livingstone-Banks and Ian Hesketh for proofreading my work. Finally, a thank-you to Dan Carey for providing me with an opportunity to present my research at Galway ain 2014.

1. Susan Sheets-Pyenson, *John William Dawson: Faith, Hope and Science* (Montreal: McGill-Queen's University Press, 1996), 123.
2. John William Dawson, "Review of Darwin, *On the Origin of Species by Means of Natural Selection*," *Canadian Naturalist and Geologist* 5 (1860): 100–120.
3. Sheets-Pyenson, *John William Dawson*; Suzanne Zeller, *Inventing Canada: Early Victorian Science and the Idea of a Transcontinental Nation*, Carleton Library (Montreal: McGill-Queen's University Press, 2009), 214; Charles F. O'Brien, *Sir William Dawson, a Life in Science and Religion* (Philadelphia: American Philosophical Society, 1971).
4. Bernard Lightman, "Scientists as Materialists in the Periodical Press: Tyndall's Belfast Address," in *Science Serialized: Representations of the Sciences in Nineteenth-Century Periodicals*, ed. Geoffrey Cantor and Sally Shuttleworth (Cambridge: MIT Press, 2004), 199–237; Bernard Lightman, *Victorian Science in Context* (Chicago: University of Chicago Press, 1997); Peter Harrison, "'Science' and 'Religion': Constructing the Boundaries," *Journal of Religion* 86, no. 1 (1 January 2006): 81–106; Crosbie Smith, *The Science of Energy: A Cultural History of Energy Physics in Victorian Britain* (Chicago: University of Chicago Press, 1998); Jona-

than Topham, "Beyond the 'Common Context': The Production and Reading of the Bridgewater Treatises," *Isis* 89, no. 2 (1998): 233–62; Jonathan Topham, "Science and Popular Education in the 1830s: The Role of the 'Bridgewater Treatises,'" *British Journal for the History of Science* 25, no. 4 (1992): 397–430; John Hedley Brooke and G. N. Cantor, *Reconstructing Nature: The Engagement of Science and Religion* (Oxford: Oxford University Press, 1998); Frank M. Turner, "The Victorian Conflict between Science and Religion: A Professional Dimension," *Isis* 69 (1978): 356–76.

5. Aileen Fyfe, *Science and Salvation: Evangelical Popular Science Publishing in Victorian Britain* (Chicago: University of Chicago Press, 2004); Bernard Lightman, *Victorian Popularizers of Science: Designing Nature for New Audiences* (Chicago: University of Chicago Press, 2007).
6. Ralph O'Connor, *The Earth on Show: Fossils and the Poetics of Popular Science, 1802–1856* (Chicago: University of Chicago Press, 2007); James Secord, "Introduction," in *Principles of Geology [Selections from 1830–33],* by Charles Lyell (London: Penguin Classics, 1997), ix–xliii; Ronald L. Numbers, *The Creationists* (Berkeley: University of California Press, 1993); David Livingstone, *Adam's Ancestors: Race, Religion, and the Politics of Human Origins* (Baltimore: Johns Hopkins University Press, 2008).
7. John William Dawson, *Archaia: Or, Studies of the Cosmogony and Natural History of the Hebrew Scriptures* (Montreal: B. Dawson & Son, 1860), 215.
8. Peter R. Eakins and Jean S. Eakins, "Sir John William Dawson," available at *Dictionary of Canadian Biography Online*, http://www.biographi.ca/en/bio/dawson_john_william_12E.html/.
9. Dawson, *Archaia*, preface.
10. George Fetherling, *The Rise of the Canadian Newspaper* (Oxford: Oxford University Press, 1990), 58–59, 78–79.
11. Patricia Fleming, Yvan Lamonde, and Fiona Black, eds., *History of the Book in Canada: 1840–1918* (Toronto: University of Toronto Press, 2005), 186–87.
12. Fleming, Lamonde, and Black, *History of the Book*, 301.
13. Sheets-Pyenson, *John William Dawson*, 122.
14. Sheets-Pyenson, *John William Dawson*, 123–24.
15. Anonymous, "Archaia; Or, Studies of the Cosmogony and Natural History of the Hebrew Scriptures," *Athenaeum* 1711 (11 August 1860): 198.
16. Dawson, *Archaia*, 17.

17. Fyfe, *Science and Salvation*, 4.

18. Dawson, *Archaia*, 20.

19. Dawson, *Archaia*, 75.

20. Dawson, *Archaia*, 317.

21. See, for example, Lightman, *Victorian Popularizers of Science*; Lightman, *Victorian Science in Context*; Aileen Fyfe and Bernard Lightman, *Science in the Marketplace: Nineteenth-Century Sites and Experiences* (Chicago: University of Chicago Press, 2007); Fyfe, *Science and Salvation*.

22. Numbers, *The Creationists*, 141.

23. Zeller, *Inventing Canada*, 6.

24. Dawson, *Archaia*, 100.

25. Numbers, *The Creationists*. 21.

26. Dawson, *Archaia*, 102.

27. For more on the day-age theory, see Numbers, *The Creationists*.

28. Dawson, *Archaia*, 98–99.

29. For more on Hugh Miller see, for example, Michael A. Taylor, *Hugh Miller: Stonemason, Geologist, Writer* (Edinburgh: NMS Enterprises, 2007); Hugh Miller, *Hugh Miller's Memoir: From Stonemason to Geologist* (Edinburgh: Edinburgh University Press, 1995); David Livingstone and Charles W. J. Withers, *Geographies of Nineteenth-Century Science* (Chicago: University of Chicago Press, 2011); O'Connor, *The Earth on Show*.

30. O'Connor, *The Earth on Show*, 397.

31. Dawson, *Archaia*, 113.

32. Dawson, *Archaia*, 84.

33. Dawson, *Archaia*, 327.

34. Secord, "Introduction," xvi; O'Connor, *The Earth on Show*.

35. Secord, "Introduction," xvii–xxiii.

36. Secord, "Introduction," xxx.

37. Secord, "Introduction," xxiii.

38. Secord, "Introduction," xxiv.

39. Sheets-Pyenson, *John William Dawson*, 24–32.

40. Dawson, *Archaia*, 83.

41. Dawson, *Archaia*, 53, 83–85, 327.

42. Mott T. Greene, *Geology in the Nineteenth Century: Changing Views of a Changing World* (Ithaca: Cornell University Press, 1982), 22.

43. Dawson, *Archaia*, 354.

44. Dawson, *Archaia*, 116.

45. Dawson, *Archaia*, 168, 194–200, 341–44.

46. Dawson, *Archaia*, 116.

47. Dawson, *Archaia*, 208.

48. For more on prehistoric mammals see Chris Manias's chapter (this volume).

49. Dawson, *Archaia*, 116.

50. Dawson, *Archaia*, 353.

51. Dawson, *Archaia*, 219–20.

52. Dawson, *Archaia*, 222.

53. Dawson, *Archaia*, 246.

54. Sheets-Pyenson, *John William Dawson*, 190–203.

55. James A. Secord, *Victorian Sensation: The Extraordinary Publication, Reception, and Secret Authorship of* Vestiges of the Natural History of Creation (Chicago: University of Chicago Press, 2003), 37, 168–70.

56. For more on Dawson's critique of Darwin, see, for example, David Livingstone, *Dealing with Darwin: Place, Politics, and Rhetoric in Religious Engagements with Evolution* (Baltimore: Johns Hopkins University Press, 2014), 95–98; Ronald L. Numbers and John Stenhouse, eds., *Disseminating Darwinism: The Role of Place, Race, Religion, and Gender* (Cambridge: Cambridge University Press, 1999); Suzanne Zeller, "Environment, Culture, and the Reception of Darwin in Canada, 1859–1909," in Numbers and Stenhouse, *Disseminating Darwinism*, 91–122; Neil S. Forkey, *Canadians and the Natural Environment to the Twenty-First Century* (Toronto: University of Toronto Press, 2012), 28–30; Sheets-Pyenson, *John William Dawson*, 125–35.

57. Dawson, *Archaia*, 230.

58. Dawson, *Archaia*, 251–52.

59. Dawson, *Archaia*, 252.

60. Efram Sera-Shriar, *The Making of British Anthropology, 1813–1871* (London: Pickering and Chatto, 2013).

61. John Cornell, "From Creation to Evolution: Sir William Dawson and the Idea of Design in the Nineteenth Century," *Journal of the History of Biology* 16, no. 1 (1983): 137–70.

62. Dawson, *Archaia*, 253.

63. Dawson, *Archaia*, 254–55.

64. Dawson, *Archaia*, 275.

65. David N. Stamos, *Darwin and the Nature of Species* (New York: SUNY Press, 2012), 135–45; Monte Harrell Hampton, *Storm of Words: Science, Religion, and Evolution in the Civil War Era* (Tuscaloosa: University of Alabama Press, 2014), 68–71; B. Ricardo Brown, *Until Darwin, Science, Human Variety and the Origins of Race* (London: Pickering and Chatto, 2010), 89–98, 252–77.

66. Livingstone, *Adam's Ancestors*, 95.

67. Dawson, *Archaia*, 281.

68. Dawson, *Archaia*, 282.

69. Cornell, "From Creation to Evolution," 144.

70. Dawson, *Archaia*, 293.

71. Dawson, *Archaia*, 282.

72. Dawson, *Archaia*, 294–99.

73. Dawson, *Archaia*, 32.

74. Dawson, *Archaia*, 41.

75. Dawson, *Archaia*, 41.

76. Livingstone, *Adam's Ancestors*, 8–24.

77. Dawson, *Archaia*, 17.

78. Dawson, *Archaia*, 301.

79. Dawson, *Archaia*, 307.

80. Dawson, *Archaia*, 18–19.

81. Hannah Franziska Augstein, *James Cowles Prichard's Anthropology: Remaking the Science of Man in Early Nineteenth-Century Britain* (Amsterdam: Rodopi, 1999).

82. Dawson, *Archaia*, 299, 303–5.

83. Dawson, *Archaia*, 308.

84. Dawson, *Archaia*, 231, 238, 309.

85. Dawson, *Archaia*, 217, 235.

86. Dawson, *Archaia*, 237–38.

87. Dawson, *Archaia*, 237–38.

88. Dawson, *Archaia*, 307–15.

89. See also Peter C. Kjærgaard, "'Hurrah for the Missing Link!' A History of Apes, Ancestors and a Crucial Piece of Evidence," *Notes and Records: The Royal Society Journal of the History of Science* 65 (2011): 83–98; Gillian Beer, *Forging the Missing Link: Interdisciplinary Stories: Inaugural Lecture, Delivered 18 November 1991* (Cambridge: Cambridge University Press Archive, 1992); Jane Goodall, *Perfor-*

mance and Evolution in the Age of Darwin: Out of the Natural Order (Abingdon: Routledge, 2002).

90. Dawson, *Archaia*, 349.
91. Dawson, *Archaia*, 244.
92. Sheets-Pyenson, *John William Dawson*, 125.
93. Dawson, "Review."
94. John Duns, "Archaia; Or, Studies of the Cosmogony and Natural History of the Hebrew Scriptures," *North British Review* 32, no. 64 (May 1860): 548.
95. Anonymous, "Archaia," *Saturday Review of Politics, Literature, Science and Art* 9, no. 225 (18 February 1860): 212.
96. Numbers, *The Creationists*, 11.

CHAPTER 3: HISTORICIZING BELIEF

1. Edward Burnett Tylor, *Anahuac: or Mexico and the Mexicans, Ancient and Modern* (London: Longman, Green, Longman, and Roberts, 1861).
2. Edward Burnett Tylor, *Researches into the Early History of Mankind and the Development of Civilisation* (London: John Murray, 1865).
3. Edward Burnett Tylor, *Primitive Culture: Researches into the Development of Mythology, Philosophy, Religion, Language, Art and Custom*, 2 vols. (London: John Murray, 1871). Unless otherwise noted, this is the edition cited.
4. Robert Ranulph Marett, *Tylor* (London: Chapman and Hall, 1936).
5. George Stocking, *Victorian Anthropology* (New York: Free Press, 1987); Robert Kenny, "From the Curse of Ham to the Curse of Nature: The Influence of Natural Selection on the Debate on Human Unity before the Publication of *The Descent of Man*," *British Journal for the History of Science* 40 (2007): 363–88; David Livingstone, *Adam's Ancestors: Race, Religion, and the Politics of Human Origins* (Baltimore: Johns Hopkins University Press, 2008); Frank Turner, "The Victorian Conflict between Science and Religion: A Professional Dimension," *Isis* 69 (1978): 356–76; Bernard Lightman, "Victorian Sciences and Religion: Discordant Harmonies," *Osiris* 16 (2001): 343–66; Janet Browne, "A Science of Empire: British Biogeography before Darwin," *Revue d'Histoire des Sciences* 45 (1992): 453–75; Michael Bravo, "Ethnological Encounters," in *Cultures of Natural History*, ed. Nick Jardine, James Secord, and Emma Spary (Cambridge: Cambridge University Press, 1996), 338–57; Daniel Carey, "Compiling Nature's History: Travellers and Travel Narratives in the Early Royal Society," *Annals of Science* 54 (1997):

269–92; Peter Hulme and Tim Youngs, "Introduction," in *Cambridge Companion to Travel Writing*, ed. Peter Hulme and Tim Youngs (Cambridge: Cambridge University Press, 2002), 1–16.

6. For more on the scientific naturalists and religion, see Frank Turner, *Between Science and Religion: The Reaction to Scientific Naturalism in Late Victorian England* (New Haven: Yale University Press, 1974); Turner, "The Victorian Conflict"; Bernard Lightman, *The Origins of Agnosticism: Victorian Unbelief and the Limits of Knowledge* (Baltimore: Johns Hopkins University Press, 1987); Lightman, "Victorian Sciences"; Bernard Lightman, "Does the History of Science and Religion Change depending on the Narrator? Some Atheist and Agnostic Perspectives," *Science and Christian Beliefs* 24 (2012): 149–68. Recently, Timothy Larsen has also written on Tylor and his alleged crisis of faith; see Timothy Larsen, *The Slain God: Anthropologists and the Christian Faith* (Oxford: Oxford University Press, 2014).
7. For a more nuanced understanding of the boundaries between science and religion, see John Brooke and Geoffrey Cantor, "Whose Science? Whose Religion?" in *Reconstructing Nature: The Engagement of Science and Religion*, ed. John Brooke and Geoffrey Cantor (Oxford: Oxford University Press, 1998), 43–72; Peter Harrison, "'Science' and 'Religion': Constructing the Boundaries," *Journal of Religion* 86, no. 1 (2006): 81–106.
8. Ivan Strenski, *Understanding Theories of Religion: An Introduction* (Chichester: John Wiley and Sons, 2015), 45.
9. George Stocking, *After Tylor: British Social Anthropology, 1888–1951* (Madison: University of Wisconsin Press, 1995), 3–4.
10. Andrew Lang, "Edward Burnett Tylor," in *Anthropological Essays Presented to Edward Tylor in Honour of his 75th Birthday*, ed. Henry Balfour (Oxford: Clarendon Press, 1907), 1.
11. George Stocking, "Edward Burnett Tylor and the Mission of Primitive Man," in *Delimiting Anthropology: Occasional Essays and Reflections*, ed. George Stocking (Madison: University of Wisconsin Press, 2001), 105.
12. Henry James Sumner Maine, *Ancient Law: Its Connection with the Early History of Society and its Relations to Modern Ideas* (London: John Murray, 1861); John Ferguson McLennan, *Primitive Marriage: An Inquiry into the Origin of the Form of Capture in Marriage Ceremonies* (Edinburgh: Adam and Charles Black, 1865); John Lubbock, *Pre-Historic Times: As Illustrated by Ancient Remains, and the Manners and Customs of Modern Savages* (London: Williams and Norgate, 1865).

13. George Stocking, "'Cultural Darwinism' and 'Philosophical Idealism' in E. B. Tylor," in *Race, Culture, and Evolution: Essays in the History of Anthropology*, ed. George Stocking (Chicago: University of Chicago Press, 1968), 91–109.

14. Elizabeth Isichei, *Victorian Quakers* (Oxford: Oxford University Press, 1970), 7; Geoffrey Cantor, *Quakers, Jews and Science: Religious Responses to Modernity and the Sciences in Britain, 1650–1900* (Oxford: Oxford University Press, 2005), 234–36.

15. Bravo, "Ethnological Encounters," 339–40; Efram Sera-Shriar, "Arctic Observers: Richard King, Monogenism and the Historicisation of Inuit through Travel Narratives," *Studies in History and Philosophy of Biological and Biomedical Sciences* (2015): 24–25.

16. Stocking, *Victorian Anthropology*, 157.

17. Tylor, *Anahuac*, 1.

18. Stocking, *Victorian Anthropology*, 157; Efram Sera-Shriar, *The Making of British Anthropology, 1813–1871* (London: Pickering and Chatto, 2013), 154–55.

19. For more on the British in the Spanish world, see Mary Louise Pratt, *Imperial Eyes: Travel Writing and Transculturation* (Abingdon: Routledge, 1992), 15–16; Joan-Pau Rubiés, *Travel and Ethnology in the Renaissance: South India through European Eyes, 1250–1625* (Cambridge: Cambridge University Press, 2000), 2–4; N. A. M. Rodger, *The Command of the Ocean: A Naval History of Britain, 1649–1815* (London: Penguin Books, 2006), 23–26; David Rock, "The British in Argentina: From Informal Empire to Postcolonialism," in *Informal Empire in Latin America: Culture, Commerce, and Capital*, ed. Matthew Brown (Oxford: Basil Blackwell, 2008), 49–77; Brendan Simms, *Three Victories and a Defeat: The Rise and Fall of the First British Empire* (London: Penguin Books, 2008).

20. Kristine Jones, "Nineteenth Century British Travel Accounts of Argentina," *Ethnohistory* 33 (1986): 196–97; Daniela Bleichmar, *Visible Empire: Botanical Expeditions and Visual Culture in the Hispanic Enlightenment* (Chicago: University of Chicago Press, 2012), 20–25.

21. For examples see Tylor, *Anahuac*, 147.

22. Tylor, *Anahuac*, iii.

23. Sera-Shriar, *The Making of British Anthropology*, 154–57. Other scholars have also discussed the significance of Tylor's journey through Mexico as a formative period in his anthropological work. See Ivan Strenski, *Thinking about Religion: An Historical Introduction to Theories of Religion* (Oxford: Basil Blackwell, 2006),

91; David Chidester, "Darwin's Dogs: Animals, Animism, and the Problem of Religion," *Soundings: An Interdisciplinary Journal* 92 (2009): 56; Marjorie Wheeler-Barclay, *The Science of Religion in Britain, 1860–1915* (Charlottesville: University of Virginia Press, 2010), 76.

24. Tylor, *Anahuac*, 148.
25. Tylor, *Anahuac*, 289.
26. Tylor, *Anahuac*, 50–52.
27. Tylor, *Anahuac*, 211–12.
28. Tylor, *Primitive Culture*, 1:16. See also Sera-Shriar, *The Making of British Anthropology*, 168–69.
29. Strenski, *Thinking about Religion*, 91.
30. Tylor, *Anahuac*, 222–23.
31. George Stocking, "What's in a Name? The Origins of the Royal Anthropological Institute (1837–71)," *Man* 6 (1971): 369–90; Kenny, "From the Curse of Ham."
32. Stocking, *Victorian Anthropology*, 159.
33. The anthropological schism is far too complex a topic to discuss here. However, there has been a large amount of secondary literature written on subject. For example see John W. Burrow, "Evolution and Anthropology in the 1860s: The Anthropological Society of London, 1863–71," *Victorian Studies* 7 (1963): 137–49; Stocking, "What's in a Name?" 369–90; Kenny, "From the Curse of Ham"; Efram Sera-Shriar, "Observing Human Difference: James Hunt, Thomas Huxley, and Competing Disciplinary Strategies in the 1860s," *Annals of Science* 70 (2013): 461–91.
34. Edward Burnett Tylor, *Anthropology: An Introduction to the Study of Man and Civilisation* (London: Macmillan, 1881).
35. Tylor discusses the earlier versions of his theory of animism in the preface to *Primitive Culture*, 1:v.
36. Wheeler-Barclay, *The Science of Religion*, 75.
37. For a general discussion on scientific naturalism in Victorian Britain, see Gowan Dawson and Bernard Lightman, eds., *Victorian Scientific Naturalism: Community, Identity, and Continuity* (Chicago: University of Chicago Press, 2014).
38. Tylor, *Primitive Culture*, 1:2.
39. Huxley's drive to place all the scientific disciplines onto an evolutionary framework is far too big a subject for a discussion here. For more on Huxley and evolutionary theory, see Mario DiGregorio, *T. H. Huxley's Place in Natural Science*

(New Haven: Yale University Press, 1984), 53–76. For the relationships among ethnology, anthropology, and Darwinism, and for the debates between monogenists and polygenists, see Adrian Desmond and James Moore, *Darwin's Sacred Cause: Race, Slavery and the Quest for Human Origins* (London: Penguin Books, 2009), 199–227; Sera-Shriar, *The Making of British Anthropology*, 109–45.

40. For more on the debates relating to politics, religion, and human origin in the nineteenth century, see Livingstone, *Adam's Ancestors*; Martin Fichman, "Biology and Politics: Defining the Boundaries," in *Victorian Science in Context*, ed. Bernard Lightman (Chicago: University of Chicago Press, 1997), 94–101.

41. For more on Huxley's vocational strategy and scientific naturalism, see Edward Caudill, "The Bishop-Eaters: The Publicity Campaign for Darwin and on the Origin of Species," *Journal of the History of Ideas* 55 (1994): 441–60; Paul White, *Thomas Huxley: Making the "Man of Science"* (Cambridge: Cambridge University Press, 2003), 51–58; James Elwick, *Styles of Reasoning in the British Life Sciences: Shared Assumptions, 1820–58* (London: Pickering and Chatto, 2007), 131–59; Bernard Lightman, *Victorian Popularizers of Science: Designing Nature for New Audiences* (Chicago: University of Chicago Press, 2007), 359–61; Nanna Katrine Lüders Kaalund, "Oxford Serialised: Revisiting the Huxley-Wilberforce Debate through the Periodical Press," *History of Science* 52 (2014), 429–53.

42. Edward Burnett Tylor, *Primitive Culture: Researches into the Development of Mythology, Philosophy, Religion, Language, Art and Custom,* 2nd ed., 2 vols. (London: Henry Holt, 1889), 1:vii–viii.

43. George Stocking also discussed the significance of Prichardian monogenism in shaping Tylor's writings. See Stocking, *Victorian Anthropology*, 157–59.

44. James Cowles Prichard, *Researches into the Physical History of Man* (London: John and Arthur Arch, 1813), 1–2.

45. Tylor, *Primitive Culture*, 1:12, 5.

46. Tylor, *Primitive Culture*, 1:6.

47. Tylor, *Primitive Culture*, 1:1.

48. For more on the importance of travel narratives in the making of nineteenth-century ethnographic knowledge, see Stocking, *Victorian Anthropology*, 78–109; Bravo, "Ethnological Encounters"; Janet Browne, "Biogeography and Empire," in *Cultures of Natural History*, ed. Nicholas Jardine, James Secord, and Emma Spary (Cambridge: Cambridge University Press, 1996), 305–21; Sera-Shriar, *The Making of British Anthropology*, 43–51; Sera-Shriar, "Arctic Observers," 24–25.

49. For more on the importance of natural history and medicine for nineteenth-century ethnology and anthropology, see Efram Sera-Shriar, "What Is Armchair Anthropology? Observational Practices in Nineteenth-Century British Human Sciences," *History of the Human Sciences* 27 (2014): 32–33. For more on observational practices in nineteenth-century British anthropology, see Sera-Shriar, *The Making of British Anthropology*, 1–20.

50. Tylor, *Primitive Culture*, 1:8.

51. Tylor, *Primitive Culture*, 1:8–11.

52. Tylor, *Primitive Culture*, 1:18. For more on Auguste Comte's positivism, see Brooke and Cantor, "Whose Science? Whose Religion?" 47–57; Ian Hesketh, *The Science of History in Victorian Britain* (London: Pickering and Chatto, 2011), 17–19.

53. Tylor, *Primitive Culture*, 1:33. See also Laavanyan Ratnapalan, "E. B. Tylor and the Problem of Primitive Culture," *History and Anthropology* 19 (2008): 131–42.

54. Chris Manias, "The Problematic Construction of 'Palaeolithic Man': The Old Stone Age and the Difficulties of the Comparative Method, 1859–1914," *Studies in History and Philosophy of Biological and Biomedical Sciences* 51 (2015): 32–43. For more on the comparative method in nineteenth-century British sciences, see John W. Burrow, *Evolution and Society: A Study in Victorian Social Theory* (Cambridge: Cambridge University Press, 1966); Joan Leopold, *Culture in Comparative and Evolutionary Perspective: E. B. Tylor and the Making of Primitive Culture* (Berlin: Dietrich Reimer Verlag, 1980); Peter Bowler, *The Invention of Progress: The Victorians and the Past* (Oxford: Basil Blackwell, 1989).

55. Tylor, *Primitive Culture*, 1:15.

56. Tylor, *Primitive Culture*, 1:21.

57. Tylor, *Primitive Culture*, 1:21.

58. For more on the contestation for cultural authority between science and religion during the Victorian era, see Turner, "The Victorian Conflict"; Lightman, "Victorian Sciences."

59. Tylor, *Primitive Culture*, 1:20–21.

60. Tylor, *Primitive Culture*, 1:383.

61. Tylor, *Primitive Culture*, 2:329.

62. For one of the best studies on the relationships among science, missionaries, and race, see Sujit Sivasundaram, *Nature and the Godly Empire: Science and Evangelical Mission in the Pacific, 1795–1850* (Cambridge: Cambridge University Press, 2005). See also Catherine Hall, *Civilizing Subjects: Metropole and Colony in the English*

Imagination, 1830–1867 (Chicago: University of Chicago Press, 2002); Anna Johnston, *Missionary Writing and Empire, 1800–1860* (Cambridge: Cambridge University Press, 2003); Esme Cleall, *Missionary Discourses of Difference: Negotiating in the British Empire, 1840–1900* (Basingstoke: Palgrave Macmillan, 2012).

63. Tylor, *Primitive Culture*, 2:330.
64. Tylor references George Turner, *Nineteen Years in Polynesia: Missionary Life, Travel and Researches in the Islands of the Pacific* (London: John Snow, 1861), 88, 427.
65. William Ellis, *Polynesian Researches during a Residence of Nearly Six Years in the South Sea Islands*, 3 vols. (London: Fisher, Son and Jackson, 1829–1831), 2:235.
66. Tylor references Henry Callaway, *The Religious System of the Amazulu: Izinyanga Zokubula, or, Divination, as Existing among the Amazulu, in their own Words* (London: Trübner, 1868), 124, 141, 174, 182.
67. Browne, "Biogeography and Empire," 306–8.
68. Richard Taylor, *The Past and Present of New Zealand with Its Prospects for the Future* (London: William Macintosh, 1868).
69. Joel Samuel Polack, *Manners and Customs of the New Zealanders; with Notes Corroborative of their Habits, Usages Etc.*, 2 vols. (London: James Madden, 1840).
70. For more on Polack in New Zealand, see Lazarus Goldman, *The History of Jews in New Zealand* (Wellington: Reed, 1958), 34–39.
71. For the most thorough list of reviews of Tylor's *Primitive Culture*, consult Leopold, *Culture in Comparative and Evolutionary Perspective*, 180–81.
72. Anonymous, "Primitive Culture: Researches into the Development of Mythology, Philosophy, Religion, Art and Custom," *Athenaeum* 2271 (1871): 558.
73. Anonymous, "Mr Tylor on Primitive Culture," *Examiner* 3304 (1871): 536.
74. Henry Calderwood, "Primitive Culture: Researches into the Development of Mythology, Philosophy, Religion, Art, and Custom," *Contemporary Review* 19 (1871): 211.
75. Calderwood, "*Primitive Culture*," 210.
76. Anonymous, "Degradation or Development," *Calcutta Review* 108 (1872): 251, 266.
77. Anonymous, "Tylor's Primitive Culture," *North American Review* 114 (1872): 227.
78. Alfred Russel Wallace, "Primitive Culture: Researches into the Development of Mythology, Philosophy, Religion, Art, and Custom," *The Academy* 3 (1872): 69–71. For an example of Tylor on spiritualism, see Tylor, *Primitive Culture* (1871), 1:405. For more on Tylor and Wallace, see also Wheeler-Barclay, *The Science of Religion*, 87.

CHAPTER 4: THE HISTORY OF THE "RED MAN"

We would like to thank Efram Sera-Shriar for his constructive critiques and insightful observations. We also want to thank Juan Manuel Rodriguez Caso for his valuable comments. This work was supported by FONDECYT–Chile (National Fund for Scientific and Technological Development), project number 11140388.

1. Jean Franco, "Un viaje poco romántico: Viajeros británicos hacia Sudamérica, 1818–28," *Escritura, Teoría y Crítica Literaria* 4 (1979): 129–42; David Rock, "The British in Argentina: From Informal Empire to Postcolonialism," *Bulletin of Latin American Research* 24 (2008): 49–77.
2. Efram Sera-Shriar, "Tales from Patagonia: Phillip Parker King and Early Ethnographic Observation in British Ethnology, 1826–1830," *Studies in Travel Writing* 19 (2015): 204–23.
3. Mary Louise Pratt, *Imperial Eyes: Travel Writing and Transculturation* (London: Routledge, 1992), 147.
4. For biographical information, see William Bollaert, *Texas*, ed. Eugene Hollon and Ruth Butler (Norman: University of Oklahoma Press, 1989).
5. The works of Bollaert, Henry Christy, and Alfred Maudslay were important for the construction of Latin American archaeological collections in Britain; see Margarita Díaz-Andreu, *A World History of Nineteenth-Century Archaeology: Nationalism, Colonialism, and the Past* (Oxford: Oxford University Press, 2007).
6. Michael Bravo, "Ethnological Encounters," in *Cultures of Natural History*, ed. Nicholas Jardine, James Secord, and Emma Spary (Cambridge: Cambridge University Press, 1996), 338–57.
7. See Efram Sera-Shriar, *The Making of British Anthropology, 1813–1871* (London: Pickering and Chatto, 2013).
8. Sera-Shriar, "Tales from Patagonia," 204–23.
9. For more on the history of the distinction between armchair anthropology and field anthropology, see Sera-Shriar, *The Making of British Anthropology*.
10. On this argument, see Friedrich Otto Hertz, *Race and Civilization* (New Jersey: Ktav, 1970).
11. David Livingstone, *Adam's Ancestors: Race, Religion, and the Politics of Human Origins* (Baltimore: Johns Hopkins University Press, 2008), 169.
12. Bravo, "Ethnological Encounters."
13. See Anthony Pagden, *The Fall of Natural Man: The American Indian and the*

Origins of Comparative Ethnology (Cambridge, Cambridge University Press, 1982).

14. Richard Drayton, *Nature's Government: Science, Imperial Britain, and the "Improvement" of the World* (Telangana: Orient Longman, 2005), 92–93.
15. Daniel Headrick, *Power over Peoples: Technology, Environments, and Western Imperialism, 1400 to the Present* (Princeton, NJ: Princeton University Press, 2009).
16. Pagden, *The Fall of Natural Man.* Unsurprisingly, Bollaert knew and used seventeenth-century Spanish literary sources. He especially acknowledged José de Acosta's famous *Historia natural y moral de las Indias.*
17. Franco, "Un viaje poco romántico," 129.
18. Franco, "Un viaje poco romántico."
19. Domingo F. Sarmiento, *Facundo o civilización i barbarie en las pampas argentinas* (Santiago: Imprenta del Progreso, 1845), 52.
20. For a more detailed account of the Ecuador Land Company, see Robert A. Stafford, *Scientist of Empire: Sir Roderick Murchison, Scientific Exploration and Victorian Imperialism* (Cambridge: Cambridge University Press, 1989). See also Elisa Sevilla Pérez, "Imperios informales y naciones poscoloniales: La autoridad de la ciencia" (PhD thesis, Facultad Latinoamericana de Ciencias Sociales Sede Ecuador, 2011).
21. Manuel Villavicencio, *Geografía de la Republica de Ecuador* (New York: Robert Graighead, 1858), 179. Bollaert considered Villavicencio's book an important reference. He wrote, "Villavicencio's volume is a standard authority for the geography of Ecuador, and deserves to be translated." William Bollaert, *Antiquarian, Ethnological and Other Researches in New Granada, Equador, Peru and Chile, with Observations on the Pre-Incarial, Incarial, and Other Monuments of Peruvian Nations* (London: Trübner, 1860), 74.
22. Villavicencio, *Geografía*, 165.
23. Villavicencio, *Geografía*, 184.
24. Villavicencio quoted in Sevilla Pérez, "Imperios informales," 81.
25. Nancy Stepan, *The Hour of Eugenics: Race, Gender and Nation in Latin America* (Ithaca: Cornell University Press, 1991).
26. Rock, "The British in Argentina," 51.
27. William Bollaert, "Observations on the Past and Present Populations of the New World," *Memoirs Read before the Anthropological Society of London* 1 (1864): 97.
28. William Bollaert, "Contributions to an Introduction to the Anthropology of the

New World," *Memoirs Read before the Anthropological Society of London* 2 (1866): 92.

29. Sera-Shriar, *The Making of British Anthropology*, 51.
30. Johann Friedrich Blumenbach, "On the Natural Variety of Mankind," in *The Idea of Race*, ed. Robert Bernasconi and Tommy L. Lott (Indianapolis: Hackett, 2000), 29, 35.
31. For more on the influence of Blumenbach's writings on British ethnology see chapters 1 and 3 of Sera-Shriar, *The Making of British Anthropology*.
32. Bravo, "Ethnological Encounters."
33. Chris Manias, *Race, Science, and the Nation: Reconstructing the Ancient Past in Britain, France and Germany* (London: Routledge, 2013), 30.
34. Bollaert, *Antiquarian*, 3.
35. Thomas Hodgkin and Richard Cull, "A Manual of Ethnological Inquiry," *Journal of the Ethnological Society of London* 3 (1854): 193.
36. Sera-Shriar, *The Making of British Anthropology*, 85.
37. Dr. Hodgkin cofounded the Aborigines Protection Society (APS) in 1837. See Jack Morrell and Arnold Thackray, *Gentlemen of Science: Early Years of the British Association for the Advancement of Science* (Oxford: Oxford University Press, 1981), 284. The APS had a philanthropic vein and prioritized the protection and preservation of extra-European populations in the empire. See also Sera-Shriar, *The Making of British Anthropology*.
38. Hodgkin and Cull, "A Manual of Ethnological Inquiry," 193.
39. Bollaert, *Antiquarian*, 195; Efram Sera-Shriar, "Ethnology in the Metropole: Robert Knox, Robert Gordon Latham and Local Sites of Observational Training," *Studies in History and Philosophy of Biological and Biomedical Sciences* 42 (2011): 491.
40. Bollaert, *Antiquarian*, 492.
41. Bollaert, "Contributions," 150.
42. Robert E. Bieder, *Science Encounters the Indian, 1820–1880: The Early Years of American Ethnology* (Norman: University of Oklahoma Press, 1986); George Stocking, *Victorian Anthropology* (New York: Free Press, 1987); Robert Kenny, "From the Curse of Ham to the Curse of Nature: The Influence of Natural Selection on the Debate on Human Unity before the Publication of *The Descent of Man*," *British Journal for the History of Science* 40 (2007): 363–88; Sera-Shriar, *The Making of British Anthropology*.

43. Bollaert, *Antiquarian*; William Bollaert, "Introduction to the Paleography of America; Or, Observations on Ancient Picture and Figurative Writing in the New World," *Memoirs Read before the Anthropological Society of London* 1 (1864): 169–94; William Bollaert, "Some Accounts of the Astronomy of the Red Man of the New World," *Memoirs Read before the Anthropological Society of London* 1 (1864): 210–80; Bollaert, "Contributions."
44. See Bollaert, *Antiquarian.*
45. Bollaert, "Observations," 72.
46. Bollaert, "Contributions," 150.
47. Bollaert, "Contributions," 93. See also Pagden, *The Fall of Natural Man.*
48. Bollaert, "Contributions," 131.
49. Manias, *Race, Science, and the Nation*, 26.
50. Bollaert, "Contributions," 150.
51. Livingstone, *Adam's Ancestors*, 110.
52. Bollaert, "Contributions," 94–95.
53. Bollaert, "Some Accounts," 210. See also Bollaert, "Introduction to the Paleography," 169.
54. Bollaert, "Some Accounts," 278.
55. Bollaert, "Some Accounts," 279 .
56. George C. Lewis, "The Astronomy of the Ancients," *Ethnological Journal: A Monthly Record of Ethnological Research and Criticism* (1865): 211–21.
57. Lewis, "The Astronomy," 217.
58. See Anne Fabian, *The Skull Collectors: Race, Science and America's Unburied Dead* (Chicago: University of Chicago Press, 2010), 16.
59. Bollaert, "Observations," 114.
60. Bollaert, "Contributions," 151.
61. Kenny, "From the Curse of Ham," 372.
62. See Livingstone, *Adam's Ancestors.*
63. Bollaert, "Contributions," 151.
64. Ephraim George Squier, *Notes on Central America* (New York: Harper and Brothers, 1855), 60, 111.
65. Bollaert, *Antiquarian*, 223.
66. Pagden, *The Fall of Natural Man.*
67. Bollaert, "Observations," 80.
68. Bollaert, "Observations," 78.

69. Bollaert, "Observations," 85.
70. Bollaert, "Observations," 87.
71. Bollaert, "Observations," 90.
72. Bollaert, "Contributions," 150.
73. See Juan Manuel Rodríguez Caso, "Anthropology in Transition: A Study of the Sciences of Man at the British Association for the Advancement of Science, 1866–1870" (PhD diss., University of Leeds, 2014).
74. Kenny, "From the Curse of Ham," 382.

CHAPTER 5: HISTORICIZING HUMANS IN COLONIAL INDIA

I am very grateful to James Poskett, Efram Sera-Shriar, and Sujit Sivasundaram for their perceptive comments on earlier drafts of this chapter. I would also like to thank Pratik Chakrabarti for an insightful conversation on geology and theories of race in colonial India. The epigraph is from Herbert Hope Risley, *The People of India*, ed. W. Crooke (Calcutta: Thacker, Spink, 1915), 4.

1. Risley, *The People of India*, 4.
2. Herbert Hope Risley, "The Study of Ethnology in India," *Journal of the Anthropological Institute of Great Britain and Ireland* 20 (1891): 235–63.
3. Susan Bayly, *Caste, Society and Politics in India from the Eighteenth Century to the Modern Age* (Cambridge: Cambridge University Press, 1999), 127.
4. On British institutions devoted to human variation studies, see George Stocking, *Victorian Anthropology* (New York: Free Press, 1987); Efram Sera-Shriar, *The Making of British Anthropology, 1813–1871* (London: Pickering and Chatto, 2013).
5. Bernard Cohn, *Colonialism and Its Forms of Knowledge: The British in India* (Princeton, NJ: Princeton University Press, 1996), 16–56.
6. Partha Chatterjee, *Nationalist Thought and the Colonial World: A Derivative Discourse?* (London: Zed Books, 1986).
7. The diffusion model of scientific knowledge was elucidated most famously in George Basalla, "The Spread of Western Science," *Science* 156 (1967): 611–22. For a critique of Basalla in relation to colonial India, see David Arnold, *Science, Technology and Medicine in Colonial India* (Cambridge: Cambridge University Press, 2000), 9–18. For an overview of work critical of diffusionism, see Roy MacLeod, "Introduction: Nature and Empire: Science and the Colonial Enterprise," *Osiris* 15 (2000): 1–13. The most significant recent theoretical moves beyond Basalla

include James Secord, "Knowledge in Transit," *Isis* 95 (2004): 654–72; Simon Schaffer et al., eds., *The Brokered World: Go-Betweens and Global Intelligence, 1770–1820* (Sagamore Beach: Watson, 2009); Kapil Raj, "Beyond Postcolonialism . . . and Postpositivism: Circulation and the Global History of Science," *Isis* 104 (2013): 337–47.

8. On science and globality, see Sujit Sivasundaram, ed., "Global Histories of Science" Focus Section, *Isis* 101 (2010): 95–158.
9. Jean Grondin, "Gadamer's Basic Understanding of Understanding," in *The Cambridge Companion to Gadamer*, ed. Robert J. Dostal (Cambridge: Cambridge University Press, 2002), 36–51.
10. William Jones, *The Works of Sir William Jones. In Six Volumes*, ed. Anna Maria Shipley Jones (London: G. G. and J. Robinson, 1799), 6:295.
11. Jones, *Works*, 1:22.
12. Jones, *Works*, 1:295.
13. Jones, *Works*, 1:311–12.
14. Jones, *Works*, 1:281–82.
15. Jones, *Works*, 1:134–35.
16. Kapil Raj, "Mapping Knowledge Go-Betweens in Calcutta, 1770–1820," in Schaffer et al., *The Brokered World*, 105–50.
17. Joan-Pau Rubiés, "New Worlds and Renaissance Ethnology," *History and Anthropology* 6, no. 2 (1993): 157–97; Siep Stuurman, "François Bernier and the Invention of Racial Classification," *History Workshop Journal* 50 (Autumn 2000): 1–21; David Livingstone, *Adam's Ancestors: Race, Religion, and the Politics of Human Origins* (Baltimore: Johns Hopkins University Press, 2008), 42.
18. Jones, *Works*, 1:34, 92–93.
19. Jones, *Works*, 1:26. Jones had posited an "almost primeval" single ancestor of Persian, Greek, Latin, and the Celtic languages in 1779 before arriving in India. See Garland Cannon, *The Life and Mind of Oriental Jones: Sir William Jones, the Father of Modern Linguistics* (Cambridge: Cambridge University Press, 1990), 243.
20. Thomas Trautmann, *Aryans and British India* (Berkeley: University of California Press, 1997), 45–47; Jones, *Works*, 1:20.
21. Jones, *Works*, 1:130–31; Trautmann, *Aryans*, 42.
22. Tony Ballantyne, *Orientalism and Race: Aryanism in the British Empire* (Basingstoke: Palgrave Macmillan, 2002), 32–41.

23. James Cowles Prichard, *Researches into the Physical History of Man* (London: John and Arthur Arch, 1813), 438, 470–72.

24. Prichard, *Researches into the Physical History of Man*, 433–35.

25. James Cowles Prichard, *Researches into the Physical History of Mankind*, 3rd ed. (London: Sherwood, Gilbert, and Piper, 1847), 5:98–99, 125–26, 570.

26. Stocking, *Victorian Anthropology*, 8–45; Sera-Shriar, *The Making of British Anthropology*, 21–52.

27. Trautmann, *Aryans*, 165–89.

28. Jones, *Works*, 1:28.

29. Simon Schaffer, "The Asiatic Enlightenments of British Astronomy," in Schaffer et al., *The Brokered World*, 51. For more on Henry Maine, see Ballantyne, *Orientalism and Race*, 51–52; Karuna Mantena, *Alibis of Empires: Henry Maine and the Ends of Liberal Imperialism* (Princeton, NJ: Princeton University Press, 2010).

30. Mantena, *Alibis.*

31. Prichard's adoption of Jones's postulated Indo-European descent from Ham was highly controversial. See Ballantyne, *Orientalism and Race*, 39.

32. Jones, *Works*, 1:118–20.

33. Trautmann, *Aryans*, 146–48.

34. Schaffer, "Asiatic Enlightenments"; Anna Winterbottom, "Medicine and Botany in the Making of Madras, 1680–1720," in *The East India Company and the Natural World*, ed. Vinita Damodaran, Anna Winterbottom, and Alan Lester (Basingstoke: Palgrave Macmillan, 2014), 35–57; Matthew Edney, *Mapping an Empire: The Geographical Construction of British India, 1765–1843* (Chicago: University of Chicago Press, 1997), 165–95.

35. Francis W. Ellis, "Note to the Introduction," in A. D. Campbell, *A Grammar of the Teloogoo Language, commonly termed the Gentoo, peculiar to the Hindoos inhabiting the north eastern provinces of the Indian Peninsula* (Madras: College Press, 1816), 2; see also xix–xxv, 3.

36. Robert Caldwell, *A Comparative Grammar of the Dravidian or South-Indian Family of Languages* (London: Harrison, 1856), 69.

37. Trautmann, *Aryans*, 151–52.

38. Alexander Hamilton, "Review of Francis Buchanan's 'A Journey from Madras,'" *Edinburgh Review* (October 1808): 93; Trautmann, *Aryans*, 148–49.

39. John Stevenson, "Observations on the Marathi Language," *Journal of the Royal Asiatic Society of Great Britain and Ireland* 7 (1843): 84–91.

40. Trautmann, *Aryans*, 155–56.

41. Sumit Guha, "Lower Strata, Older Races, and Aboriginal Peoples: Racial Anthropology and Mythical History Past and Present," *Journal of Asian Studies* 57 (1998): 424.

42. Guha, "Lower Strata," 423–41; Sujit Sivasundaram, "Imperial Transgressions: The Animal and Human in the Idea of Race," *Comparative Studies of South Asia, Africa and the Middle East*, 35 (2015): 156–72.

43. Martin Gaenszle, "Brian Hodgson as Ethnographer and Ethnologist," and David Arnold, "Hodgson, Hooker and the Himalayan Frontier," in *The Origins of Himalayan Studies: Brian Houghton Hodgson in Nepal and Darjeeling 1820–1858*, ed. David M. Waterhouse (Abingdon: Routledge, 2004), 206–26, 189–205.

44. Trautmann, *Aryans*, 159–60; Gaenszle, "Brian Hodgson," 207–13.

45. Hodgson, quoted in Gaenszle, "Brian Hodgson," 220.

46. Michael S. Dodson, *Orientalism, Empire, and National Culture: India, 1770–1880* (Basingstoke: Palgrave Macmillan, 2007), 80–81.

47. Thomas Simpson, "Colonial Frontiers in North-Western and North-Eastern British India during the Nineteenth Century" (PhD diss., University of Cambridge, 2015), 220–65.

48. Gaenszle, "Brian Hodgson," 222; Trautmann, *Aryans*, 145, 161–64.

49. Prichard, *Physical History of Mankind*, 4:140–44.

50. Prichard, *Physical History of Mankind*, 4:244–45.

51. Prichard, *Physical History of Mankind*, 4:248–49.

52. Prichard, *Physical History of Mankind*, 4:217–18. Prichard referred to the Kalash, following his sources Mountstuart Elphinstone and Alexander Burnes, as "Kafirs" and "Siah-Posh."

53. T. Postans, "On the Biluchi Tribes Inhabiting Sindh in the Lower Valley of the Indus and Cutchi," *Journal of the Ethnological Society of London* 1 (1848): 103.

54. C. A. Bayly, *Empire and Information: Intelligence Gathering and Social Communication in India, 1780–1870* (Cambridge: Cambridge University Press, 1996), 97–141; Thomas Simpson, "'Clean out of the Map': Knowing and Doubting Space at India's High Imperial Frontiers," *History of Science* 55, no. 1 (2017): 3–36.

55. John Bryan Neufville, "On the Geography and Population of Asam," *Asiatick Researches* 16 (1828): 331–52.

56. Neufville, "On the Geography," 338.

57. Neufville, "On the Geography," 339–40.

58. On the colonial focus on slavery, see Mandy Sadan, *Being and Becoming Kachin: Histories beyond the State in the Borderworlds of Burma* (Oxford: Oxford University Press, 2013), 50–70.
59. Neufville, "On the Geography," 339–40.
60. Neufville, "On the Geography," 348–50.
61. Edward Tuite Dalton, *Descriptive Ethnology of Bengal* (Calcutta: Office of the Superintendent of Government Printing, 1872), 9–13.
62. Lynn Zastoupil, *Rammohun Roy and the Making of Victorian Britain* (Basingstoke: Palgrave Macmillan, 2010), 4.
63. Prichard, *Physical History of Mankind*, 4:237n.
64. Prichard, *Physical History of Mankind*, 4:244.
65. Zastoupil, *Rammohun Roy*, 41–43.
66. The watercolor rendering of Savinghac's print is Anonymous, *Portrait, head and shoulders, of Raja Ram Mohan Roy* (London: British Library, Oriental and India Office Collections, Prints & Drawings, WD1288, c.1832). Other portraits produced during Rammohan's time in Britain showing him with fairer skin include Henry Perronet Briggs, *Portrait of Rammohun Roy*, oil on canvas (Bristol: Bristol City Museum and Art Gallery, 1832); Rembrandt Peale, *The Rajah Rammahun Roy* (Salem, MA: Peabody Essex Museum, 1833). See also Zastoupil, *Rammohun Roy*, 48–53.
67. Anonymous, "On the Life, Character, and Cerebral Development of Rajah Rammohun Roy," *Phrenological Journal and Miscellany* 8 (1834): 591.
68. Anon., "Rajah Rammohun Roy," 577.
69. Anon., "Rajah Rammohun Roy," 591–92.
70. Stocking, *Victorian Anthropology*, 46–77.
71. Shruti Kapila, "Race Matters: Orientalism and Religion, India and Beyond c.1770–1880," *Modern Asian Studies* 41 (2007): 471–513; Anonymous, "Catalogue of Casts of Skulls of Different Nations," *Phrenological Journal and Miscellany* 6 (1829–1830): 144; James Poskett, "Metropole of the Mind: Phrenology and the Making of a Global Science, 1815–1860" (PhD diss., University of Cambridge, 2015), 241–66.
72. Trautmann, *Aryans*, 53–54.
73. Poskett, "Metropole of the Mind," 241–66.
74. David L. Gosling, "Darwin and the Hindu Tradition: 'Does What Goes around Come Around?'" *Zygon* 46 (2011): 345–69.

75. Kapila, *Race Matters.*

76. Tapan Raychaudhuri, *Europe Reconsidered: Perceptions of the West in Nineteenth-Century Bengal* (New Delhi: Oxford University Press, 2002), 8. See also Ballantyne, *Orientalism and Race*, 44; Trautmann, *Aryans*, 183–85, 195–96, 219–20.

77. Marwa Elshakry, *Reading Darwin in Arabic* (Chicago: University of Chicago Press, 2013), 6–7.

78. Robert Kenny, "From the Curse of Ham to the Curse of Nature: The Influence of Natural Selection on the Debate on Human Unity before the Publication of *The Descent of Man*," *British Journal of the History of Science* 40 (2007): 367.

79. Cheever Mackenzie Brown, *Hindu Perspectives on Evolution: Darwin, Dharma, and Design* (Abingdon: Routledge, 2012), 63; C. A. Bayly, *Recovering Liberties: Indian Thought in the Age of Liberalism and Empire* (Cambridge: Cambridge University Press, 2012), 252.

80. Pratik Chakrabarti, *Western Science in Modern India: Metropolitan Methods, Colonial Practices* (Delhi: Permanent Black, 2004), 175; Deepak Kumar, "Reason, Science and Religion: Gleanings from the Colonial Past," *Current Science* 99 (2010): 677.

81. Elshakry, *Reading Darwin in Arabic*, 4–5; Thomas F. Glick, ed., *The Comparative Reception of Darwin* (Chicago: University of Chicago Press, 1988).

82. Shruti Kapila, "Self, Spencer and Swaraj: Nationalist Thought and Critiques of Liberalism, 1890–1920," *Modern Intellectual History* 4 (2007): 109–27; Bayly, *Recovering Liberties*, 253.

83. Quoted in Brown, *Hindu Perspectives on Evolution*, 108, 112.

84. Brown, *Hindu Perspectives on Evolution*, 6–7.

85. C. Singh, *Life and Teachings of Swami Dayanand Saraswati* (New Delhi: Jan Gyan Prakashan, 1971), pt. 2, 5; Brown, *Hindu Perspectives on Evolution*, 123–24.

86. Chatterjee, *Nationalist Thought*, 67–69; Gyan Prakash, *Another Reason: Science and the Imagination of Modern India* (Princeton, NJ: Princeton University Press, 1999), 57–58; Raychaudhuri, *Europe Reconsidered*, 145–46; Bankim Chandra Chatterjee, "Mill, Darwin, and Hinduism," *Bankimchandra Chatterjee: Sociological Essays*, ed. S. N. Mukerjee and M. Maddern (Calcutta: Rddhi, 1986), 60–70.

87. Quoted in Chatterjee, *Nationalist Thought*, 69.

88. Swami Vivekananda, "Evolution," in *The Complete Works*, by Swami Vivekananda

(Calcutta: Advaita Ashrama, 1973), 5:277–79; Kenny, "From the Curse of Ham," 383–84.

89. Brown, *Hindu Perspectives on Evolution*, 78, 139–40.

90. Swami Vivekananda, "The East and West," in Vivekananda, *Complete Works*, 5:519.

91. Keshab Chandra Sen, "Philosophy and Madness in Religion," *Keshab Chandra Sen and the Brahma Samaj*, ed. T. E. Slater (Madras: Society for Promoting Christian Knowledge, 1884), 87.

92. Sen, "Philosophy and Madness," 87–88.

93. Gosling, "Darwin and the Hindu Tradition," 350–55; Brown, *Hindu Perspectives on Evolution*, 131–72.

94. Aurobindo Ghose, *Evolution* (Calcutta: Arya, 1944), 2.

95. Bayly, *Recovering Liberties*, 254–56; Brown, *Hindu Perspectives on Evolution*, 133–34.

96. Michael Bravo, *The Accuracy of Ethnoscience: A Study of Inuit Cartography and Cross-Cultural Commensurability* (Manchester: Manchester Papers in Social Anthropology, 1996), 6–13; Helen Tilley, "Global Histories, Vernacular Science, and African Genealogies; or, Is the History of Science Ready for the World?" *Isis* 101 (2010): 110–19.

CHAPTER 6: HOW AND WHY DARWIN GOT EMOTIONAL ABOUT RACE

This chapter is dedicated to Jonathan Hodge. Many thanks to Efram Sera-Shriar for inviting this chapter; to Jonathan Hodge for much-appreciated help and encouragement along the way; and to audiences in Halifax (Canada), Cambridge, and London for valuable discussion of an earlier version.

1. Charles Darwin, *The Descent of Man, and Selection in Relation to Sex*, 2 vols. (London: John Murray, 1871), 1:73. Scans of all books by Darwin cited here can be accessed at the Darwin Online website, http://darwin-online.org.uk/.

2. Charles Darwin, *The Expression of the Emotions in Man and Animals* (London: John Murray, 1872), 365.

3. On the *Expression*, the best introduction remains Janet Browne, "Darwin and the Expression of the Emotions," in *The Darwinian Heritage*, ed. David Kohn (Princeton, NJ: Princeton University Press) , 307–26. A very stimulating recent collection of papers on the book is Angelique Richardson, ed., *After Darwin: Animals, Emotions, and the Mind* (New York: Rodopi, 2013).

4. Darwin, *Expression of the Emotions*, 246–49.

5. Darwin, *Expression of the Emotions*, 361.

6. Darwin, *Descent of Man*, 1:228–35. See also Stephen Alter, "Race, Language, and Mental Evolution in Darwin's *Descent of Man*," *Journal of the History of the Behavioral Sciences* 43 (2007): 239–55.

7. Gregory Radick, "Darwin's Puzzling *Expression*," *Comptes Rendus Biologies* 333 (2010): 181–87, offers a critical summary of this subliterature along with a preliminary sketch of the argument of this essay.

8. Darwin, *Descent of Man*, 1:152–53.

9. On the *Expression* and Darwin's Lamarckism, see Gregory Radick, "Darwin on Language and Selection," *Selection* 3 (2002): 11–14.

10. Alan Fridlund, "Darwin's Anti-Darwinism in the *Expression of the Emotions in Man and Animals*," in *International Review of Studies on Emotion*, ed. K. T. Strongman (Wiley: Chichester, 1992), 2:117–37.

11. Darwin, *Descent of Man*, 1:107–36; Gregory Radick, "Darwin and Humans," in *The Cambridge Encyclopedia of Darwin and Evolutionary Thought*, ed. Michael Ruse (Cambridge: Cambridge University Press, 2013), 175–77.

12. See, e.g., Fridlund, "Darwin's Anti-Darwinism," 125, 128–30; Paul Ekman, "Introduction (and Commentaries)," in *The Expression of the Emotions in Man and Animals*, by Charles Darwin, 3rd ed. (London: HarperCollins, 1998), xxxiii–iv, along with many of his subsequent remarks interpolated into Darwin's text; and Steven Pinker's review of Ekman's edition of the *Expression*, "Still Stimulating after All These Years," *Science* 281 (24 July 1998): 522–23.

13. For further discussion see Radick, "Darwin's Puzzling *Expression*," 182–83.

14. Darwin, *Expression of the Emotions*, 361.

15. Elliott Sober, who calls the inference from similarity to common ancestry "modus Darwin," provides an exhaustive analysis of the difficulties besetting it (there are many more than I touch on here) in Elliott Sober, *Evidence and Evolution: The Logic behind the Science* (Cambridge: Cambridge University Press, 2008). For discussion, see Gregory Radick, "Evidence-Based Darwinism," *Biological Theory* 5 (2010): 289–91.

16. Darwin, *Expression of the Emotions*, 361, 147.

17. Darwin, *Expression of the Emotions*, 361–64.

18. Darwin, *Expression of the Emotions*, 147–77, 362.

19. Darwin, *Expression of the Emotions*, 363, drawing on 226–28; Darwin, *Descent of Man*, 1:140–44.

20. Darwin, *Expression of the Emotions*, 337–47, 364.

21. Darwin, *Expression of the Emotions*, 364–65.

22. On later famous Darwinian ideas originating after the notebook period as being in the minority, see M. J. S. Hodge, "The Notebook Programmes and Projects of Darwin's London Years," in *The Cambridge Companion to Darwin*, 2nd ed., ed. Jonathan Hodge and Gregory Radick (Cambridge: Cambridge University Press, 2009), 44. On 1866 as the birth year (or near to it) of Darwin's expression questionnaire, see Paul White, "Darwin's Emotions: The Scientific Self and the Sentiment of Objectivity," *Isis* 100 (2009): 817. The 1867 volume of the Darwin *Correspondence* has a useful appendix on the questionnaire and associated activity that year: Frederick Burkhardt et al., eds., *The Correspondence of Charles Darwin*, 30 vols. (Cambridge: Cambridge University Press, 1985–), 15:525–26.

23. Scans of the M and N notebooks together with transcriptions can be accessed at the Darwin Online website, http://darwin-online.org.uk/. The "industry standard" printed transcriptions with annotations are in Paul H. Barrett et al., eds., *Charles Darwin's Notebooks, 1836–1844* (Cambridge: Cambridge University Press, 1987), 517–96. But an older volume, well worth seeking out for its thoughtful and informative commentary, and including key excerpts from Darwin's other notebooks, is Paul H. Barrett, *Metaphysics, Materialism and the Evolution of Mind: Early Writings of Charles Darwin* (Chicago: University of Chicago Press, 1980), itself an abbreviated edition of Howard E. Gruber, *Darwin on Man: A Psychological Study of Creativity* (London: Wildwood House, 1974).

24. Here I adopt the usual convention of giving the notebook name followed by the page number.

25. An apparent exception is an entry on blushing at N15 that begins: "Does a negress blush[?]" But it is immediately answered in the affirmative, with Darwin recording his near certainty that the Fuegians that he traveled with on the *Beagle* blushed. The entry carries on to consider animal blushing, the physiology of blushing, and the blushing of children.

26. On Darwin's notebook theorizing on expression as looking back to *Zoonomia* rather than forward to the *Origin of Species* and the *Descent of Man*, see William Montgomery, "Charles Darwin's Thought on Expressive Mechanisms in Evolu-

tion," in *The Development of Expressive Behavior: Biology-Environment Interactions*, ed. Gail Zivin (London: Academic Press, 1985), 38–42.

27. Gregory Radick, *Edinburgh and Darwin's* Expression of the Emotions (Edinburgh: Institute for Advanced Studies in the Humanities, 2009).

28. Darwin to Lyell, 10 January 1860, in Burkhardt et al., *Correspondence*, 8:28. All letters cited here can also be found online at the Darwin Correspondence Project website, https://www.darwinproject.ac.uk/.

29. Burkhardt et al., *Correspondence*, 8:19–21; Richard B. Freeman and Peter J. Gautrey, "Charles Darwin's *Queries about Expression*," *Bulletin of the British Museum (Natural History)*, Historical Series 4 (1972): 213–14.

30. Burkhardt et al., *Correspondence*, 8:400–401, 401.

31. On Bridges, Sulivan, and Darwin, see Gregory Radick, "Did Darwin Change His Mind about the Fuegians?" *Endeavour* 34 (2010): 50–54.

32. On Stirling's visit, see Henry Norton Sulivan, ed., *Life and Letters of the Late Admiral Sir Bartholomew James Sulivan, K.C.B., 1810–1890* (London: John Murray, 1896), 387.

33. Sulivan to Darwin, 11 January 1867, in Burkhardt et al., *Correspondence*, 15:25.

34. Sulivan to Darwin, 25 December 1866, Darwin Correspondence Project, "Letter no. 5325," accessed on 4 August 2016, at http://www.darwinproject.ac.uk/DCP-LETT-5325/.

35. Darwin to Sulivan, 31 December 1866, Darwin Correspondence Project, "Letter no. 5330," accessed on 4 August 2016, at http://www.darwinproject.ac.uk/DCP-LETT-5330/.

36. Sulivan to Darwin, 11 January 1867, in Burkhardt et al., *Correspondence*, 15:25.

37. Darwin to Sulivan, 15 January 1867, in Burkhardt et al., *Correspondence*, 15:35. An exchange of letters between Sulivan and Darwin in 1881 about Bridges's work on the language of the Fuegians went on to have a remarkable afterlife within Christian missionary polemics about Darwin's supposed change of mind at the end of his life about the lowliness of the Fuegians; see Radick, "Did Darwin Change?" 50–54.

38. Darwin wrote to T. H. Huxley on 7 January about sending "the M.S of my Big book . . . to the Printers," and of now "thinking of a Chapter on man." Darwin to Huxley, 7 January 1867, in Burkhardt et al., *Correspondence*, 15:15–16, 15.

39. See the letters from Darwin to Fritz Müller (22 February 1867), Julius von Haast (27 February 1867), Robert Swinhoe (27 February 1867), J. P. M. Weale (27 Feb-

ruary 1867), and Ferdinand von Mueller (28 February 1867), and from Asa Gray to Darwin (26 March 1867), in Burkhardt et al., *Correspondence*, 15:92–94, 110–16, 175–76. On the *Queries about Expression*, see, in addition to the dedicated appendix in this *Correspondence* volume, 525–26, and Freeman and Gautrey, "Charles Darwin's *Queries* [1972]," a brief follow-up paper by them, "Charles Darwin's *Queries about Expression*," *Journal of the Society for the Bibliography of Natural History* 7 (1975). Both papers, along with five versions of the printed queries, are available at the Darwin Online website, http://darwin-online.org.uk/.

40. On Brooke, see the letter from Wallace to Darwin, in Burkhardt et al., *Correspondence,* 15:119–20; Darwin, *Expression of the Emotions*, 21. On Swinhoe, see note 6 on the letter from Darwin to P. L. Slater, 4 May 1861, Darwin Correspondence Project, "Letter no. 3138," accessed on 4 August 2016, http://www.darwinproject.ac.uk/DCP-LETT-3138; Darwin, *Expression of the Emotions*, 21.

41. I take the phrase "information order" from Simon Schaffer, "Newton on the Beach: The Information Order of *Principia Mathematica*," *History of Science* 47 (2009): 243–76. (Schaffer took it from C. A. Bayly.)

42. Darwin, *Expression of the Emotions*, 15.

43. In his M notebook, Darwin not only noted the salutary effect of cold water on him but speculated boldly on its significance. "It is," he wrote, "an argument for materialism that cold water brings on suddenly in head, a frame of mind, analogous to those feelings, which may be considered as truly spiritual" (M19).

44. See Mike Dixon and Gregory Radick, *Darwin in Ilkley* (Stroud: History Press, 2009), 26, for the quotation, from Darwin to Lyell, 30 September 1859, in Burkhardt et al., *Correspondence*, 7:338.

45. Charles Darwin, *On the Origin of Species by Means of Natural Selection* (London: John Murray, 1859), 19–20.

46. Darwin, *Origin of Species*, 253–54.

47. On this correspondence in context, see Adrian Desmond and James Moore, *Darwin's Sacred Cause: Race, Slavery and the Quest for Human Origins* (London: Penguin Books, 2009), 311–13; Dixon and Radick, *Darwin in Ilkley*, 79–83. Darwin returned to dogs at much greater length in Charles Darwin, *The Variation of Animals and Plants Under Domestication*, 2 vols. (London: John Murray, 1868), where he explained that "the main argument in favour of the several breeds of the dog being the descendants of distinct wild stocks, is their resemblance in various countries to distinct species still existing there" (1:19–20). The most exten-

sive discussion of Darwin's views on the origins of dogs can be found in Clare O'Reilly, "Darwin, Dogs and the Tree of Life: A Revisionist Account of the Role of Cross-Breeding and Hybridism in Darwin's Evolutionary Theorising" (MA diss., University of Leeds, 2014). Scientific debate on the evolutionary origins of dogs remains lively, though there is consensus that they derive, as Lyell thought, from ancestral wolves. See Gemma Tarlach, "The Origins of Dogs," *Discover* (December 2016): 32–39.

48. Lyell to Darwin, 22 October 1859, in Burkhardt et al., *Correspondence*, 13:418–19 (emphasis added).

49. Darwin to Lyell, 25 October 1859, in Burkhardt et al., *Correspondence*, 7:357 (emphasis in original).

50. Lyell to Darwin, 28 October 1859, in Burkhardt et al., *Correspondence*, 7:363. See also, in the same volume, Darwin to Lyell, 31 October 1859 (363–64), and Lyell to Darwin, 21 November 1859 (384–85). Darwin referred to this correspondence in a letter sent around the same time to his older sister, Caroline, who shared Lyell's worries about the Pallasian dog passage ("Lyell was bothered on same point and I have not expressed myself clearly. By my theory, all dogs . . . have descended from some one very ancient species" [386–87, 386]).

51. Darwin to Lyell, 23 November 1859, in Burkhardt et al., *Correspondence*, 7:391–93 (emphasis in original).

52. Darwin, *Descent of Man*, 1:229.

53. On Darwin's work in Ilkley on what became the second edition of the *Origin of Species*, see Dixon and Radick, *Darwin in Ilkley*, 107–9.

54. Darwin, *Origin of Species*, 427.

55. See M. J. S. Hodge, "Darwin, the Galápagos and His Changing Thoughts about Species Origins, 1835–1837," *Proceedings of the California Academy of Sciences* 61 (2010).

56. The cover of the M notebook reads: "This Book full of Metaphysics on Morals and Speculations on Expression—1838 [/] Selected Dec. 16, 1856." On the cover of the N notebook Darwin wrote something similar, with the same date.

57. I do not mean to suggest that the suitability of Darwin's notebook theorizing on expression for making a common-ancestry case was the *only* attraction it held for him. When Lamarckian use inheritance was widely accepted, Lamarckian explanations for behavior could seem, compared with natural-selection explanations, both better evidenced (because based on habit formation, which anyone could

witness, rather than on chance inheritable brain variations, which no one could witness) and more powerful (because where natural selection can explain in a general way why, say, emotions are expressed at all, and even why pairs of expressions come to be the opposite of one another, typically it cannot explain why particular expressions express the emotions they do—which is put down to chance). Even into the 1940s, evolutionarily minded biologists who were otherwise well disposed toward natural selection found Lamarckian explanations of the origins of behavior much more straightforwardly intelligible—hence the considerable relief they felt when the Baldwin Effect, where adaptive habitual behavior leads and selection follows, emerged as a theoretical option. See, on these themes, Radick, "Darwin on Language," 12–14; and Gregory Radick, "Animal Agency in the Age of the Modern Synthesis: W. H. Thorpe's Example," in *Animal Agents: The Non-human in the History of Science*, ed. Amanda Rees, British Journal for the History of Science Themes 2 (Cambridge: Cambridge University Press, 2017), 35–56.

58. The subtitle of the American edition is *How a Hatred of Slavery Shaped Darwin's Views on Human Evolution*. Unconvinced reviewers include Robert J. Richards, "Descent of Man," *American Scientist* 97 (2009), and Patricia Fara, "[Review of Desmond and Moore, *Darwin's Sacred Cause*]," *Journal of Historical Geography* 35 (2009): 614–16. Cf. Gregory Radick, "[Review of Desmond and Moore, *Darwin's Sacred Cause*]," *Times Higher Education*, 12 February 2009, 48–49.

59. Darwin, *Descent of Man*, 1:217–29; Darwin, *Origin of Species*, 419–20.

60. On the recurrence of surprise-bonus argument structures, at every scale, throughout Darwin's published writings, and on their link to an ideal of scientific argumentation known as the *vera causa* ideal, see Gregory Radick, "Race and Language in the Darwinian Tradition (and What Darwin's Language–Species Parallels Have to Do with It)," *Studies in History and Philosophy of Biological and Biomedical Sciences* 39 (2008): 363.

61. Desmond and Moore, *Darwin's Sacred Cause: Race, Slavery and the Quest for Human Origins*, esp. chaps. 1 and 3.

62. Darwin to Fox, 4 September 1850, Darwin Correspondence Project, "Letter no. 1352," accessed on 4 August 2016, at http://www.darwinproject.ac.uk/DCP-LETT-1352/. Desmond and Moore misidentify the recipient as Lyell; Desmond and Moore, *Darwin's Sacred Cause*, 242.

63. Darwin, *Origin of Species*, 23.

64. John Bachman, *The Doctrine of the Unity of the Human Race Examined on the*

Principles of Science (Charleston, SC: Canning, 1850), 8 (and 135–36 on pigeons). On Bachman, see Desmond and Moore, *Darwin's Sacred Cause*, esp. 176, 194, 209–14.

65. Desmond and Moore, *Darwin's Sacred Cause*, esp. chaps. 8 and 9, with a pithy summary on 243–44.
66. The author traveler was Frederick Law Olmsted, better remembered today as the designer of Central Park in New York City. On Darwin as a reader of Olmsted's, see Desmond and Moore, *Darwin's Sacred Cause*, 308–9.
67. Charles Darwin, *Journal of Researches into the Natural History and Geology of the Countries Visited During the Voyage of H.M.S. Beagle Round the World*, 2nd ed. (London: John Murray, 1845), 500. For discussion, see Desmond and Moore, *Darwin's Sacred Cause*, 180–83. Later in life Darwin learned with pleasure that the American abolitionist William Lloyd Garrison had admired the added passage; Desmond and Moore, *Darwin's Sacred Cause*, xx.
68. Desmond and Moore, *Darwin's Sacred Cause*, xvii.
69. Darwin, *Expression of the Emotions*, 234–35.
70. For further discussion on Darwin and race, see Radick, "Did Darwin Change?" and Radick, "Race and Language" (with light criticism of Desmond and Moore on this score in Radick, "[Review]"). On Lincoln and race, see Garry Wills, "Lincoln's Black History," *New York Review of Books* (11 June 2009).
71. On empire and the *Beagle*, see Hodge, "Notebook Programmes," 70; on Darwin, Fitzroy, and the missionaries, see Radick, "Did Darwin Change?" 53.
72. It should be stressed that, while a historian looking back can see these dependencies and inversions, Darwin betrayed not even the slightest awareness of them.
73. Ekman, "Introduction (and Commentaries)," 24.
74. Hong-Jin Liu, "Data and the Development of Research Methods in the Science of Human Emotional Expression from Darwin to Klineberg" (PhD diss., University of Leeds, 2016), esp. chap. 2. Well after the social psychologist Otto Klineberg (1899–1992), human expressive diversity remains a topic of interest. See, e.g., Michael Price, "Facial Expressions—Including Fear—May Not Be as Universal as We Thought," *Science* (17 October 2016), http://www.sciencemag.org/news/2016/10/facial-expressions-including-fear-may-not-be-universal-we-thought.
75. Darwin, *Expression of the Emotions*, 365–66.
76. [Anonymous], "[Review of Darwin, *Expression of the Emotions*]," *Times*, 13 December 1872, quoting from Darwin, *Expression of the Emotions*, 12 (I have

given the version in Darwin's text). A scan of the original review is available at the Darwin Online website, http://darwin-online.org.uk/. On the reviews of the *Expression* generally, see Angelique Richardson, "'The Book of the Season': The Conception and Reception of Darwin's *Expression*," in Richardson, *After Darwin*, 69–79.

77. The classic discussion of chance versus providence in Darwin's thought is M. J. S. Hodge, "Natural Selection as a Causal, Empirical, and Probabilistic Theory," in M. J. S. Hodge, *Before and after Darwin: Origins, Species, Cosmogonies, and Ontologies* (Aldershot: Ashgate, 2008), 240–46. For his more recent reflections see M. J. S. Hodge, "Chance and Chances in Darwin's Early Theorizing and in Darwinian Theory Today," in *Chance in Evolution*, ed. Grant Ramsey and Charles H. Pence (Chicago: University of Chicago Press, 2016) , 41–75.
78. See Hodge, "Notebook Programmes," 69–71; M. J. S. Hodge, "Capitalist Contexts for Darwinian Theory: Land, Finance, Industry and Empire," *Journal of the History of Biology* 42 (2009): 407–8.
79. On Lyell's conversion or, as he joked to Darwin, perversion, and the role in it of Darwin's observations about how rapidly European species had naturalized in newly colonized countries, see Dixon and Radick, *Darwin in Ilkley*, 28–29, 32n15.

CHAPTER 7: THE COMPARATIVE METHOD IN "SHALLOW TIME"

I am very grateful for the discussions with my former supervisors, Rosemary Mitchell, Nathan Uglow, and Richard Salmon, that developed the ideas underpinning this chapter. Particular thanks are due to Rosemary Mitchell for introducing me to Walter Scott. To Efram Sera-Shriar's suggestions of reading material, I owe this "history of science turn."

1. [Thomas Babington Macaulay], "The Romance of History," *Edinburgh Review* 47, no. 94 (May 1828): 333.
2. L. P. Hartley, *The Go-Between* (Harmondsworth: Penguin, 1958), 7.
3. Anne McClintock, *Imperial Leather: Race, Gender, and Sexuality in the Colonial Contest* (Abingdon: Routledge, 1995), 36, 40. For more on Tylor's concept of "survivals" see Efram Sera-Shriar's chapter (this volume).
4. Walter Scott, *Ivanhoe*, ed. Graham Tulloch (London: Penguin, 2000), 6.
5. See A. B. Van Riper, *Men among the Mammoths: Victorian Science and the Discovery of Human Prehistory* (Chicago: University of Chicago Press, 1993); James

Secord, *Victorian Sensation: The Extraordinary Publication, Reception, and Secret Authorship of* Vestiges of the Natural History of Creation (Chicago: University of Chicago Press, 2000); Peter Kjærgaard, "'Hurrah for the Missing Link!' A History of Apes, Ancestors and a Crucial Piece of Evidence," *Notes and Records: The Royal Society Journal of the History of Science* 65 (2011): 83–98.

6. Dugald Stewart, *Biographical Memoir of Adam Smith* (New York: Augustus Kelley, 1966), 33–34.
7. Ronald L. Meek, *Social Science and the Ignoble Savage* (Cambridge: Cambridge University Press, 1976), 141. See as an example volume 1 of William Robertson, *The History of America*, 3 vols. (Dublin: Messrs Whitestone, 1777–1796).
8. Roxann Wheeler, *The Complexion of Race: Categories of Difference in Eighteenth-Century British Culture* (Philadelphia: University of Pennsylvania, 2000), 37; Colin Kidd, *The Forging of Races: Race and Scripture in the Protestant Atlantic World, 1600–2000* (Cambridge: Cambridge University Press, 2006), 25.
9. Kidd, *The Forging of Races*, 121, 122. For more on the relationship between science and religion in the nineteenth century, see Frank Turner, *Between Science and Religion: The Reaction to Scientific Naturalism in Late Victorian England* (New Haven: Yale University Press, 1974); Frank Turner, "The Victorian Conflict between Science and Religion: A Professional Dimension," *Isis* 69 (1978): 356–76; Bernard Lightman, *The Origins of Agnosticism: Victorian Unbelief and the Limits of Knowledge* (Baltimore: Johns Hopkins University Press, 1987); Bernard Lightman, "Victorian Sciences and Religion: Discordant Harmonies," *Osiris* 16 (2001): 343–66; Bernard Lightman, "Does the History of Science and Religion Change depending on the Narrator? Some Atheist and Agnostic Perspectives," *Science and Christian Beliefs* 24 (2012): 149–68.
10. See David Livingstone, *Adam's Ancestors: Race, Religion, and the Politics of Human Origins* (Baltimore: Johns Hopkins University Press, 2008); Efram Sera-Shriar, *The Making of British Anthropology, 1813–1871* (London: Pickering and Chatto, 2013); Chris Manias, "The Problematic Construction of 'Palaeolithic Man': The Old Stone Age and the Difficulties of the Comparative Method, 1859–1914," *Studies in History and Philosophy of Biological and Biomedical Sciences* 51 (2015): 32–43.
11. Tony Bennett, *Pasts beyond Memory: Evolution, Museums, Colonialism* (London: Routledge, 2004), 3.
12. As Theodore Koditschek has described it, nineteenth-century writers narrated

different "histories of progress that would either keep it *to* themselves, demand it *for* themselves, or establish more stringent conditions on which it might be slowly granted to colonial others." Theodore Koditschek, *Liberalism, Imperialism, and the Historical Imagination: Nineteenth-Century Visions of a Greater Britain* (Cambridge: Cambridge University Press, 2011), 8. See also Henrika Kuklick, *The Savage Within: The Social History of British Anthropology, 1885–1945* (Cambridge: Cambridge University Press, 1991); Catherine Hall, *Civilizing Subjects: Metropole and Colony in the English Imagination, 1830–1867* (Chicago: University of Chicago Press, 2002); Andrew Thompson, *The Empire Strikes Back? The Impact of Imperialism on Britain from the Mid-Nineteenth Century* (London: Pearson Longman, 2005); Sujit Sivasundaram, *Nature and the Godly Empire: Science and Evangelical Mission in the Pacific, 1795–1850* (Cambridge: Cambridge University Press, 2005); Chris Manias, *Race, Science, and the Nation: Reconstructing the Ancient Past in Britain, France and Germany* (London: Routledge, 2013).

13. John D. Rosenberg, *Carlyle and the Burden of History* (Oxford: Oxford University Press, 1985), 9.
14. [Macaulay], "The Romance of History," 362.
15. John W. Burrow, "Images of Time: From Carlylean Vulcanism to Sedimentary Gradualism," in *History, Religion and Culture: British Intellectual History, 1750–1950*, ed. Stefan Collini, Richard Whatmore, and Brian Young (Cambridge: Cambridge University Press, 2000), 215.
16. Burrow, "Images of Time," 199.
17. [Thomas Babington Macaulay], "Ecclesiastical and Political History of the Popes of Rome during the Sixteenth and Seventeenth Centuries," *Edinburgh Review* 72 (1840): 228.
18. John W. Burrow, *Evolution and Society: A Study in Victorian Social Theory* (Cambridge: Cambridge University Press, 1966), 55.
19. Stephen Jay Gould, *Time's Arrow, Time's Cycle: Myth and Metaphor in the Discovery of Geological Time* (Cambridge, MA: Harvard University Press, 1987); Peter Bowler, *The Invention of Progress: The Victorians and the Past* (Oxford: Basil Blackwell, 1989), 9.
20. Meek, *Social Science and the Ignoble Savage*, 127.
21. See Bowler, *The Invention of Progress*, 23.
22. Bowler, *The Invention of Progress*, 90. See also Secord, *Victorian Sensation*.
23. James Macpherson, *The Poems of Ossian* (London: Cadell and Davies, 1807).

24. Peter Bowler, *The Eclipse of Darwinism: Anti-Darwinian Evolution Theories in the Decades around 1900* (Baltimore: Johns Hopkins University Press, 1983); Peter Bowler, *The Non-Darwinian Revolution: Reinterpreting a Historical Myth* (Baltimore: Johns Hopkins University Press, 1988); Peter Bowler, *Darwin Deleted: Imagining a World without Darwin* (Chicago: University of Chicago Press, 2013).
25. See Stephen Jay Gould, *Ontogeny and Phylogeny* (Cambridge, MA: Harvard University Press, 1977); and Bowler, *The Invention of Progress.*
26. Walter Scott, *Waverley; or, 'Tis Sixty Years Since*, ed. Claire Lamont (Oxford: Oxford University Press, 1986), 340.
27. See Burrow, *Evolution and Society*, 14.
28. Scott, *Ivanhoe*, 6–7.
29. Spencer Walpole, *A History of England from the Conclusion of the Great War in 1815*, 6 vols. (London: Longmans, Green, 1890), 1:22–23.
30. H. C. G. Matthew, "Walpole, Sir Spencer (1839–1907)," *Oxford Dictionary of National Biography* (Oxford: Oxford University Press, 2004), at http://www.oxforddnb.com/view/article/36712/.
31. M. M. Bakhtin, "The Bildungsroman and Its Significance in the History of Realism," in *Speech Genres and Other Late Essays*, by M. M. Bakhtin, ed. Caryl Emerson (Austin: University of Texas Press, 1986), 53; Franco Moretti, *Atlas of the European Novel, 1800–1900* (New York: Verso, 1999), 38.
32. See Meek, *Social Science and the Ignoble Savage.*
33. Scott, *Waverley*, 20.
34. Scott, *Waverley*, 72–73.
35. Edward Burnett Tylor, *Primitive Culture: Researches into the Development of Mythology, Philosophy, Religion, Language, Art and Custom*, 2 vols. (London: John Murray, 1871). See also Sera-Shriar's chapter (this volume).
36. Hall, *Civilizing Subjects*, 70.
37. William Hazlitt, *Spirit of the Age; or, Contemporary Portraits* (Plymouth: Northcote House, 1990), 102–3 (my italics).
38. [Macaulay], "The Romance of History," 363.
39. Elizabeth Gaskell, "The Well of Pen-Morfa," in *The Works of Gaskell*, ed. A. W. Ward, 2 vols. (London: Smith, Elder, 1906), 2:243.
40. Burrow, "Images of Time," 218.
41. [Thomas Carlyle], "Thoughts on History," *Fraser's Magazine* 2 (1830): 414.

42. David Amigoni, ed., *Life Writing and Victorian Culture* (Aldershot: Ashgate, 2006), 19.
43. On Carlyle as inspiration for biographies of obscure figures, see Juliette Atkinson, *Victorian Biography Reconsidered: A Study of Nineteenth-Century "Hidden" Lives* (Oxford: Oxford University Press, 2010), 9.
44. [Thomas Carlyle], *Critical and Miscellaneous Essays*, ed. Henry Duff Traill, 4 vols. (London: Chapman & Hall, 1899), 4:348.
45. Carlyle, *Critical and Miscellaneous Essays*, 350.
46. Hall, *Civilizing Subjects*, 48. For more on race science and the subjugation of Africans, see Efram Sera-Shriar, "Observing Human Difference: James Hunt, Thomas Huxley, and Competing Disciplinary Strategies in the 1860s," *Annals of Science* 70 (2013): 461–91.
47. Catherine Gallagher, *The Industrial Reformation of English Fiction: Social Discourses and Narrative Form, 1832–1867* (Chicago: University of Chicago Press, 1985), 3–35.
48. Carlyle, *Critical and Miscellaneous Essays*, 356.
49. Carlyle, *Critical and Miscellaneous Essays*, 352.
50. See Robert Knox, *The Races of Man: A Fragment* (London: Henry Renshaw, 1850). See also Efram Sera-Shriar, "Ethnology in the Metropole: Robert Knox, Robert Gordon Latham and Local Sites of Observational Training," *Studies in History and Philosophy of Biological and Biomedical Sciences* 42 (2011): 486–96.
51. Mary Midgley, *Animals and Why They Matter* (Atlanta: University of Georgia Press, 1984), 72.
52. Carlyle, *Critical and Miscellaneous Essays*, 375.
53. Peter Scheckner, ed., *An Anthology of Chartist Poetry* (London: Associated University Presses, 1989), 266.
54. Scheckner, *Chartist Poetry*, 276.
55. Caroline Norton, cited in Elizabeth Gaskell, *Mary Barton*, ed. Jennifer Foster (Peterborough: Broadview, 2000), 551.
56. Hall, *Civilizing Subjects*, 73.
57. See A.L., "The Slaves' Address to British Females" (1838), in Scheckner, *Chartist Poetry*, 82.
58. Scheckner, *Chartist Poetry*, 84.
59. Carlyle, *Critical and Miscellaneous Essays*, 353, 378, 380.
60. Midgley, *Animals and Why They Matter*, 74.

61. Julie Dugger, "Black Ireland's Race: Thomas Carlyle and the Young Ireland Movement," *Victorian Studies* 48 (2006): 467.

62. Thomas Carlyle, *Reminiscences of My Irish Journey in 1849* (London: Gilbert & Rivington, 1882).

63. Dugger, "Black Ireland's Race," 462.

64. Thomas Carlyle, *Past and Present* (London: Chapman & Hall, 1843), 201–2.

65. Ruth Schwartz Cowan, "Galton, Sir Francis (1822–1911)," *Oxford Dictionary of National Biography* (Oxford: Oxford University Press, 2005). See also D. W. Forrest, *Francis Galton: The Life and Work of a Victorian Genius* (London: Taplinger, 1974); Nicholas Wright Gillham, *A Life of Sir Francis Galton: From African Exploration to the Birth of Eugenics* (Oxford: Clarendon Press, 2001).

66. Francis Galton, *Hereditary Genius: an Inquiry into Its Laws and Consequences* (London: Macmillan, 1869), 350.

67. Emel Aileen Gökyiğit, "Reception of Francis Galton's 'Hereditary Genius' in the Victorian Periodical Press," *Journal of the History of Biology* 27 (1994): 219–21.

68. Galton, *Hereditary Genius* (1869), 343.

69. Galton, *Hereditary Genius* (1869), 336.

70. Galton, *Hereditary Genius* (1869), 340.

71. Galton, *Hereditary Genius* (1869), 4.

72. Galton, *Hereditary Genius*, rev. ed. (London: Macmillan, 1892), viii.

73. Galton, *Hereditary Genius* (1892), xii.

74. Galton, *Hereditary Genius* (1869), 5.

75. [Herman Merivale], "Galton on Hereditary Genius," *Edinburgh Review* 132 (1870): 57; Anonymous, "Hereditary Talent," *Chamber's Journal of Popular Literature* 7 (1870): 119.

76. Anonymous, *Men of the Time: A Biographical Dictionary of Eminent Living Characters of Both Sexes* (London: Routledge, 1865).

77. Galton, *Hereditary Genius* (1869), 51.

78. Sherrin Berezowsky, "Statistical Criticism and the Eminent Man in Francis Galton's *Hereditary Genius*," *Victorian Literature and Culture* 43 (2015): 831.

79. Francis Galton, *Finger Prints* (London: Macmillan, 1892), 193, 18.

80. Galton, *Finger Prints*, 196.

81. Kidd, *The Forging of Races*, 3–4.

82. Galton, *Finger Prints*, 18.

83. Galton, *Hereditary Genius* (1869), 1.

84. Carlyle, *Critical and Miscellaneous Essays*, 350–51.

85. Francis Galton, "Hereditary Talent and Character," *Macmillan's Magazine* 12 (1865): 166.

86. Bowler, *The Invention of Progress*, 11.

87. Galton, *Hereditary Genius* (1869), 345.

88. Defensive assessments include Karl Pearson, *The Life, Letters and Labours of Francis Galton*, 3 vols. (Cambridge: Cambridge University Press, 1914); Milo Keynes, ed., *Sir Francis Galton, FRS: The Legacy of His Ideas* (Basingstoke: Palgrave Macmillan, 1993); Gillham, *Sir Francis Galton*; William Revelle, "Sir Francis Galton," in *The Encyclopedia of Clinical Psychology*, ed. Robin L. Cautin and Scott O. Lilienfeld (Chichester: Wiley Blackwell, 2015), 3:1305–8; Michael Bulmer, *Francis Galton: Pioneer of Heredity and Biometry* (Baltimore: Johns Hopkins University Press, 2003). Those in condemnation include G. K. Chesterton, *Eugenics and Other Evils* (London: Cassell, 1922); Lyndsay Andrew Farrall, *The Origins and Growth of the English Eugenics Movement, 1865–1925* (London: Garland, 1985); Daniel J. Kevles, *In the Name of Eugenics: Genetics and the Uses of Human Heredity* (New York: Knopf, 1985); Diane B. Paul, *The Politics of Heredity: Essays on Eugenics, Biomedicine, and the Nature-Nurture Debate* (Albany: State University of New York Press, 1998).

89. Carlyle, *Past and Present*, 205.

90. Burrow, *Evolution and Society*, 240.

CHAPTER 8: THE FUTURE EVOLUTION OF "MAN"

I would like to thank Peter Harrison, Bernard Lightman, and Efram Sera-Shriar, as well as my colleagues at the Institute for Advanced Studies in the Humanities at the University of Queensland for their valuable critical commentary. An early version of this chapter was presented at the annual conference of the British Society for the History of Science in Swansea, UK, July 2015.

1. Charles Darwin, *The Descent of Man, and Selection in Relation to Sex* (London: John Murray, 1871), 2:405. "Man" is in quotes here to flag the fact that this was the term typically used by Victorian evolutionary writers such as Darwin when they refer to the human species.

2. On many of the social, political, and cultural issues confronting the development of evolution as a respectable scientific theory in the Victorian period, see Piers

Hale, *Political Descent: Malthus, Mutualism, and the Politics of Evolution in Victorian England* (Chicago: University of Chicago Press, 2014), and Gowan Dawson, *Darwin, Literature and Victorian Respectability* (Cambridge: Cambridge University Press, 2007).

3. See Charles Darwin to Charles Lyell, 10 December 1859, Darwin Correspondence Database, at http://www.darwinproject.ac.uk/entry-2575/ (accessed 3 December 2015).

4. Peter Bowler, *The Non-Darwinian Revolution: Reinterpreting a Historical Myth* (Baltimore: Johns Hopkins University Press, 1988); and, more recently, Peter Bowler, *Darwin Deleted: Imagining a World without Darwin* (Chicago: University of Chicago Press, 2013). On the reception of natural selection within the periodical press, see Alvar Ellegård, *Darwin and the General Reader: The Reception of Darwin's Theory of Evolution in the British Periodical Press, 1859–1872* (Chicago: University of Chicago Press, 1990).

5. For critiques of the "conflict thesis," see James Moore, *The Post-Darwinian Controversies: A Study of the Protestant Struggle to Come to Terms with Darwin in Great Britain and America, 1879–1900* (Cambridge: Cambridge University Press, 1979); John Hedley Brooke, *Science and Religion: Some Historical Perspectives* (Cambridge: Cambridge University Press, 1991); David C. Lindberg and Ronald L. Numbers, eds., *When Science and Christianity Meet* (Chicago: University of Chicago Press, 2003); Ronald L. Numbers, ed., *Galileo Goes to Jail and Other Myths about Science and Religion* (Cambridge, MA: Harvard University Press, 2009). The creation of the conflict thesis in the nineteenth century is often attributed to two books: John William Draper, *History of the Conflict between Religion and Science* (New York: D. Appleton, 1874); Andrew Dickson White, *A History of the Warfare of Science with Theology in Christendom*, 2 vols. (New York: D. Appleton, 1896).

6. Frank Turner, "The Victorian Conflict between Science and Religion: A Professional Dimension," *Isis* 69 (1978): 356–76.

7. For a recent study of the diversity of the scientific naturalists, see Gowan Dawson and Bernard Lightman, eds., *Victorian Scientific Naturalism: Community, Identity, Continuity* (Chicago: University of Chicago Press, 2014).

8. See, in particular, James Secord, *Victorian Sensation: The Extraordinary Publication, Reception, and Secret Authorship of* Vestiges of the Natural History of Creation (Chicago: University of Chicago Press, 2000); Bernard Lightman, *Victorian*

Popularizers of Science: Designing Nature for New Audiences (Chicago: University of Chicago Press, 2007); Bernard Lightman, "Darwin and the Popularization of Evolution," *Notes and Records of the Royal Society of London* 64, no. 1 (2010): 5–24.

9. Many others could have been added to this list—most notably, Herbert Spencer and H. G. Wells—but the aim here is not to be comprehensive. Much has been written about the respective cosmic evolutionisms of Spencer and Wells, whereas the key figures discussed here have received much less attention. On Spencer, see Bernard Lightman, ed., *Global Spencerism: The Communication and Appropriation of a British Evolutionist* (Leiden: Brill, 2015); on Wells see Justin Busch, *The Utopian Vision of H. G. Wells* (Jefferson, NC: McFarland, 2009).
10. For two competing perspectives on how race informed Darwin's early views of human evolution, see Adrian Desmond and James Moore, *Darwin's Sacred Cause: How a Hatred of Slavery Shaped Darwin's Views on Human Evolution* (Boston: Houghton Mifflin Harcourt, 2009), and Evelleen Richards, *Darwin and the Making of Sexual Selection* (Chicago: University of Chicago Press, 2017).
11. For a discussion of Lyell's use of this phrase and his concerns with human evolution, see John van Wyhe and Peter Kjægaard, "Going the Whole Orang: Darwin, Wallace and the Natural History of Orangutans," *Studies in History and Philosophy of Biology and Biomedical Sciences* 51 (2015): 54. See also Leonard G. Wilson, *Sir Charles Lyell's Scientific Journals on the Species Question* (New Haven: Yale University Press, 1970), 165, 262.
12. On the theological debates that were engendered by the challenges to the Mosaic narrative posed by geological, biological, and ethnological discoveries, see David Livingstone, *Adam's Ancestors: Race, Religion, and the Politics of Human Origins* (Baltimore: Johns Hopkins University Press, 2008), esp. chs. 4–6.
13. The scholarship on Wallace is vast. Particularly useful in the context of this chapter is Martin Fichman, *An Elusive Victorian: The Evolution of Alfred Russel Wallace* (Chicago: University of Chicago Press, 2004). Fichman, quite persuasively in my view, considers the various strands of Wallace's thinking, whether scientific, political, or religious, as products of an underlying and consistent teleological evolutionary cosmology.
14. Alfred Russel Wallace, "The Origin of Human Races and the Antiquity of Man Deduced from the Theory of 'Natural Selection,'" *Journal of the Anthropological Society of London* 2 (1864): clxvi.
15. Wallace, "Origin of Human Races," clxvii.

16. Wallace, "Origin of Human Races," clxviii.
17. Wallace, "Origin of Human Races," clxviii.
18. Wallace, "Origin of Human Races," clxviii.
19. Wallace, "Origin of Human Races," clxix–clxx.
20. For the discussion that was recorded at the close of Wallace's paper, see Wallace, "Origin of Human Races," clxx–clxxxvii.
21. On the importance of comparative anatomy in the ASL, see John W. Burrow, "Evolution and Anthropology in the 1860s: The Anthropological Society of London, 1863–1871," *Victorian Studies* (1963): 137–49. For a comparison of the vocational strategies of Hunt and Huxley as they related to the founding of the ASL, see Efram Sera-Shriar, "Observing Human Difference: James Hunt, Thomas Huxley, and Competing Disciplinary Strategies in the 1860s," *Annals of Science* 70 (2013): 461–91.
22. James Hunt quoted in Wallace, "Origin of Human Races," clxxviii.
23. Darwin to Wallace, 28 March [1864], Darwin Correspondence Database, http://www.darwinproject.ac.uk/entry-4510/ (accessed 9 October 2015).
24. Secord, *Victorian Sensation*.
25. See, for instance, Herbert Spencer, "Progress: Its Law and Cause," *Westminster Review* 67, no. 132 (April 1857): 445–85.
26. Wallace, "Origin of Human Races," clxx, footnote. For the influence of *Vestiges* on Wallace, see Fichman, *An Elusive Victorian*, 75–76.
27. For a more detailed analysis of Reade's evolutionary narrative, see Ian Hesketh, "A Good Darwinian? Winwood Reade and the Making of a Late Victorian Evolutionary Epic," *Studies in History and Philosophy of Biological and Biomedical Sciences* 51 (2015): 44–52.
28. Reade to Darwin, 15 September 1871, Charles Darwin Papers, number 176:48, University Library, Cambridge.
29. Reade to Darwin, 18 September 1871, Charles Darwin Papers, number 176:49, University Library, Cambridge.
30. W. Winwood Reade, *The Martyrdom of Man* (London: Trübner, 1872), v.
31. Reade, *The Martyrdom of Man*, 75.
32. Reade, *The Martyrdom of Man*, 513.
33. Darwin, *Descent of Man*, 2:404–5. For the narrative framework of *Descent of Man*, see Misia Landau, *Narratives of Evolution* (New Haven: Yale University Press, 1991), 31–60.

34. Darwin, *Descent of Man*, 2:404.

35. Darwin, *Descent of Man*, 2:403.

36. Darwin, *Descent of Man*, 2:405.

37. Greg Myers, "Nineteenth-Century Popularizations of Thermodynamics and the Rhetoric of Social Prophecy," *Victorian Studies* 29, no. 1 (1985): 36–66. For how Kelvin's theories affected the development of Darwin's thinking on evolution, see J. D. Burchfield, *Lord Kelvin and the Age of the Earth* (Chicago: University of Chicago Press, 1990).

38. Charles Darwin, *The Autobiography of Charles Darwin, 1809–1882, with Original Omissions Restored*, ed. Nora Barlow (London: Collins, 1958), 92.

39. On degenerationism and its wider political contexts, see Hale, *Political Descent*, 301–34; Daniel Pick, *Faces of Degeneration: A European Disorder, c.1848–c.1918* (Cambridge: Cambridge University Press, 1989).

40. See, for instance, Bowler, *The Non-Darwinian Revolution*, 81.

41. E. Ray Lankester, *Degeneration: A Chapter in Darwinism* (London: Macmillan, 1880), 45.

42. Lankester, *Degeneration*, 33.

43. Lankester, *Degeneration*, 212.

44. Lankester, *Degeneration*, 60–61.

45. Lankester, *Degeneration*, 61–62.

46. E. Ray Lankester, *Kingdom of Man* (New York: Henry Holt, 1907), 31.

47. Lankester, *Kingdom of Man*, 45. See also Percival Lowell, *Mars and Its Canals* (New York: Macmillan, 1906). For Lowell's embrace of a Spencerian cosmic evolutionism, see David Strauss, *Percival Lowell: The Culture and Science of a Boston Brahmin* (Cambridge, MA: Harvard University Press, 2001).

48. Lankester, *Kingdom of Man*, 41.

49. Francis Galton, *Hereditary Genius: an Inquiry into Its Laws and Consequences* (London: Macmillan, 1869), 1.

50. Francis Galton, *Inquiries into Human Faculty and Its Development* (New York: Macmillan, 1883), 23.

51. Francis Galton, "Eugenics as a Factor in Religion," in *Essays in Eugenics* (London: Eugenics Education Society, 1909), 68.

52. Galton, *Inquiries into Human Faculty*, 299–300.

53. Galton, *Inquiries into Human Faculty*, 304.

54. Galton, *Inquiries into Human Faculty*, 337.

55. Galton, *Inquiries into Human Faculty*, 301–2.

56. Janet Oppenheim, *The Other World: Spiritualism and Psychical Research in England, 1850–1914* (Cambridge: Cambridge University Press, 1985).

57. See Philip C. Almond, *Afterlife: A History of Life after Death* (London: I. B. Taurus, 2016), esp. ch. 7.

58. Frederic W. H. Myers, "Science and a Future Life," *Nineteenth Century* (April 1891): 628–29, also published in Frederic W. H. Myers, *Science and a Future Life with Other Essays* (London: Macmillan, 1893), 1–50 (quote on 3). Subsequent citations will be taken from the book.

59. Frederic W. H. Myers, "Automatic Writing," *Proceedings of the Society for Psychical Research* 3 (1885): 30, quoted in Leigh T. I. Penman, "The History of the Word *Paranormal*," *Notes and Queries* 260, no. 1 (March 2015): 32.

60. Myers, *Science and a Future Life*, 37–38.

61. Myers, *Science and a Future Life*, 38, 39.

62. Myers, *Science and a Future Life*, 49.

63. Myers, *Science and a Future Life*, 200.

64. Myers, *Science and a Future Life*, 50.

65. Myers, *Science and a Future Life*, 199.

66. The classic analysis of the Christian engagement with evolution is Moore, *The Post-Darwinian Controversies.* See also Lightman, *Victorian Popularizers of Science*, 39–94.

67. Frederick Temple, *The Relations between Religion and Science: Eight Lectures Preached before the University of Oxford in the Year 1884* (New York: Macmillan, 1884), 115.

68. St. George Jackson Mivart, *On the Genesis of Species* (London: Macmillan, 1871), 259–307. See also Livingstone, *Adam's Ancestors*, 139.

69. John Page Hopps, *The Future Life* (London: Simpkin, Marshall [1884]), 14.

70. Hopps, *The Future Life*, 16–17.

71. Hopps, *The Future Life*, 21–22.

72. Hopps, *The Future Life*, 26.

73. Hopps, *The Future Life*, 27.

74. Hopps, *The Future Life*, 155.

75. Alfred Russel Wallace, *The Scientific Aspect of the Supernatural* (London: F. Farrah, 1866), 50. See also Fichman, *An Elusive Victorian*, 175.

76. That there were limits to what natural selection could explain, particularly with

regard to human intelligence, was a theme Wallace had developed in earlier publications as well, most notably in [Alfred Russel Wallace], "Sir Charles Lyell on Geological Climates and the Origin of Species," *Quarterly Review* 126, no. 252 (1869): 359–94; Alfred Russel Wallace, "The Limits of Natural Selection as Applied to Man," in *Contributions to the Theory of Natural Selection: A Series of Essays* (London: Macmillan, 1870), 332–71.

77. Alfred Russel Wallace, *Darwinism* (London: Macmillan, 1889), 236.
78. Wallace, *Darwinism*, 236.
79. Wallace, *Darwinism*, 236.
80. Wallace, *Darwinism*, 235.
81. Wallace, *Darwinism*, 236. Wallace's views of cultural evolution were also in contrast with those of Edward Burnett Tylor. See Efram Sera-Shriar's chapter (this volume).
82. Wallace, *Darwinism*, 236.
83. Wallace, *Darwinism*, 237.
84. See Thomas Huxley, *Evolution and Ethics* (London: Macmillan, 1894); Bernard Lightman, "The 'History' of Victorian Scientific Naturalism: Huxley, Spencer, and the 'End' of Natural History," *Studies in History and Philosophy of Biological and Biomedical Sciences* 58 (2016): 17–23.
85. Michael Ruse, *Monad to Man: The Concept of Progress in Evolutionary Biology* (Cambridge, MA: Harvard University Press, 1996).
86. I discuss the influence of German Romanticism on the genre of the evolutionary epic in Hesketh, "The Recurrence of the Evolutionary Epic," *Journal of the Philosophy of History* 9, no. 2 (2015): 196–219.
87. See also Bernard Lightman, "Darwin and the Popularization of Evolution," 5–24.

AFTERWORD

1. John W. Burrow, *Evolution and Society: A Study in Victorian Social Theory* (Cambridge: Cambridge University Press, 1966); George Stocking, *Victorian Anthropology* (New York: Free Press, 1987); Roger Bannister, *Social Darwinism, Science and Myth in Anglo-American Social Thought* (Philadelphia: Temple University Press, 1979); Mike Hawkins, *Social Darwinism in European and American Thought, 1860–1945* (Cambridge: Cambridge University Press, 1997); Martin Rudwick, *Bursting the Limits of Time: The Reconstruction of Geohistory in the Age of Revolution* (Chicago: University of Chicago Press, 2005); Martin Rudwick, *Worlds before*

Adam: The Reconstruction of Geohistory in the Age of Reform (Chicago: University of Chicago Press, 2008); Peter Bowler, *Theories of Human Evolution: A Century of Debate, 1844–1944* (Baltimore: Johns Hopkins University Press, 1986); Donald Grayson, *The Establishment of Human Antiquity* (New York: Academic Press, 1983); Stephen Jay Gould, *The Mismeasure of Man* (New York: W. W. Norton, 1981); Nancy Stepan, *The Idea of Race in Science: Great Britain, 1800–1960* (London: Macmillan, 1982).

2. Robert Richards, *Darwin and the Emergence of Evolutionary Theories of Mind and Behavior* (Chicago: University of Chicago Press, 1987); Bernard Lightman, *Victorian Popularizers of Science: Designing Nature for New Audiences* (Chicago: University of Chicago Press, 2007); James Secord, *Victorian Sensation: The Extraordinary Publication, Reception, and Secret Authorship of* Vestiges of the Natural History of Creation (Chicago: University of Chicago Press, 2000); Janet Browne, *Charles Darwin: Voyaging* (Princeton, NJ: Princeton University Press, 1995); Janet Browne, *Charles Darwin: The Power of Place* (Princeton, NJ: Princeton University Press, 2002); Adrian Desmond and James Moore, *Darwin: The Life of a Tormented Evolutionist* (New York: Warner, 1991).

3. Theodore Koditschek, *Liberalism, Imperialism and the Historical Imagination: Nineteenth-Century Visions of a Greater Britain* (Cambridge: Cambridge University Press, 2011), 17, 31–55.

4. Kingstone makes a number of shrewd observations about Carlyle along the way, particularly noting that "despite a shift in his late career toward a preoccupation with more conventional and conservative hero figures, [he] remained open to appropriation as the champion of the ordinary and insignificant individual." It is, nevertheless, a long passage from his 1830 essay "On History," which declares that "social life is the aggregate of all individual men's Lives who constitute society. . . . Phoenician mariners, Italian masons, Saxon metallurgists . . . all the long forgotten train of artists and artisans; who from the first have been teaching us how to think and how to act," to the turgid, bloated volumes of the late 1850s and 1860s on Frederick the Great, which gave Hitler such comfort in his Berlin bunker of 1945. An interesting argument might be made that Sampson of Carlyle's *Past and Present* (1843) is the type, and Quashee of "The Nigger Question" (1853) the anti-type in this transition. Like all Carlylean men, each has his "work to do in the world." But why is it that Samson's natural work is to bring moral order into human organization, and Quashee's is to break his back cutting sugarcane?

Unlike the other figures studied in this volume, Carlyle never has recourse to any elaborate evolutionary theory to provide an answer. I do not think Kingstone's reference to Mary Midgley's supposed "paradox of one-way equality" sheds much light on the subject. The question remains open. See Thomas Carlyle, *Critical and Miscellaneous Essays* (London: Chapman and Hall, 1888), 2:170–71, 6:71–308; Hugh Trevor-Roper, *The Last Days of Hitler* (London: Macmillan, 1947), 87.

5. Karl Marx and Friedrich Engels, *The Communist Manifesto* (New York: International Press, 1948); Karl Marx, *Capital* (Moscow: Progress Moscow, 1954), vol. 1; Karl Marx, *A Contribution to the Critique of Political Economy* (New York: International Press, 1970).
6. Although an English translation of the *Communist Manifesto* had been available since 1850, the British evolutionists did not move in circles that were likely to encounter this text. Nevertheless, with the German publication of the first volume of *Capital* in 1867, this situation might have been expected to gradually change. In 1873 Marx presented Darwin with a copy of the second German edition, which appeared in conjunction with a French translation. Darwin acknowledged receipt but left the volume unopened and unread. Finally, with the appearance of an authorized English translation in 1886, the book began to circulate widely in the Anglo-American intellectual milieu.
7. Browne, *Darwin: The Power of Place*, 275–321. See also Adrian Desmond and James Moore, *Darwin's Sacred Cause: How a Hatred of Slavery Shaped Darwin's Views on Human Evolution* (London: Penguin Books, 2009).
8. Michael Ruse, *Monad to Man: The Concept of Progress in Evolutionary Biology* (Cambridge, MA: Harvard University Press, 1996).
9. Peter Bowler, *The Eclipse of Darwinism: Anti-Darwinian Evolution Theories in the Decades around 1900* (Baltimore: Johns Hopkins University Press, 1983); George Romanes, *Animal Intelligence* (New York: Appleton, 1891); George Romanes, *Mental Evolution in Animals* (New York: Appleton, 1894); George Romanes, *Mental Evolution in Man* (New York: Appleton, 1889); Samuel Butler, *Life and Habit* (London: Trübner, 1878). See also Samuel Butler, *Evolution, Old and New* (London: Hardwick and Bogue, 1879).
10. On Spencer, see J. D. Y. Peel, *Herbert Spencer: The Evolution of a Sociologist* (New York: Basic Books, 1971); Mark Francis and Michael W. Taylor, eds., *Herbert Spencer: Legacies* (Abingdon: Routledge, 2015); Bernard Lightman, ed., *Global Spencerism: The Communication and Appropriation of a British Evolutionist* (Lei-

den: Brill, 2015); as well as my own unpublished papers "Herbert Spencer's Late Lamarckism" (2013), and "Herbert Spencer and the 'New Liberals': Individualism, Collectivism and the Origins of the Nature/Nurture Divide" (2016), in author's possession.

11. Theodore Koditschek, "Narrative Time and Racial/Evolutionary Time in Nineteenth-Century British Liberal Imperial History," in *Race, Nation and Empire: Making Histories, 1750 to the Present*, ed. Catherine Hall and Keith McClelland (Manchester: Manchester University Press, 2010), 36–55; Koditschek, *Liberalism, Imperialism*, 206–62.
12. I am developing this argument more fully in a book manuscript in progress, tentatively titled *Race and Liberal Imperial Capitalism: Greater Britain: 1830–1914*.
13. Simpson's analysis of historical scholarship in nineteenth-century Madras needs to be supplemented by Rama Sundari Mantena, *The Origins of Modern Historiography in India: Antiquarianism and Philology, 1780–1880* (New York: Palgrave and Macmillan, 2012), who reads the dispute between Calcutta and Madras in a different way.
14. Thomas Metcalfe, *Ideologies of the Raj* (Cambridge: Cambridge University Press, 1994); Heather Streets, *The Military, Race and Masculinity in British Imperial Culture, 1857–1914* (Manchester: Manchester University Press, 2004).
15. For my own views on the position of many of the figures examined by Simpson, see Koditschek, *Liberalism, Imperialism*, 90–99, 263–313. See also Christopher Bayly, *Recovering Liberties: Indian Thought in the Age of Liberalism and Empire* (Cambridge: Cambridge University Press, 2012).
16. Partha Chatterjee, *The Nation and Its Fragments: Colonial and Postcolonial Histories* (Princeton, NJ: Princeton University Press, 1993); Sumit Sarkar, *Beyond Nationalist Frames: Postmodernism, Hindu Fundamentalism, History* (Bloomington: Indiana University Press, 2002).

BIBLIOGRAPHY

Abadía, Oscar Moro. "Thinking about the Concept of Archive: Reflections on the Historiography of Altamira." *Complutum* 24, no. 2 (2013): 145–52.

Almond, Philip C. *Afterlife: A History of Life after Death.* London: I. B. Taurus, 2016.

Alter, Stephen. "Race, Language, and Mental Evolution in Darwin's *Descent of Man.*" *Journal of the History of the Behavioral Sciences* 43 (2007): 239–55.

Amigoni, David, ed. *Life Writing and Victorian Culture.* Aldershot: Ashgate, 2006.

Anonymous. "Archaia." *Saturday Review of Politics, Literature, Science and Art* 9, no. 225 (18 February 1860): 212–13.

Anonymous. "Archaia; Or, Studies of the Cosmogony and Natural History of the Hebrew Scriptures." *Athenaeum* 1711 (11 August 1860): 198–99.

Anonymous. "Catalogue of Casts of Skulls of Different Nations." *Phrenological Journal and Miscellany* 6 (1829–1830): 144–46.

Anonymous. "Degradation or Development." *Calcutta Review* 108 (1872): 251–66.

Anonymous. "Hereditary Talent." *Chamber's Journal of Popular Literature* 7 (1870): 118–22.

Anonymous. *Men of the Time: A Biographical Dictionary of Eminent Living Characters of Both Sexes.* London: Routledge, 1865.

Anonymous. "Mr Tylor on Primitive Culture." *Examiner* 3304 (1871): 536–37.

Anonymous. "On the Life, Character, and Cerebral Development of Rajah Rammohun Roy." *Phrenological Journal and Miscellany* 8 (1834): 577–603.

Anonymous. "Primitive Culture: Researches into the Development of Mythology, Philosophy, Religion, Art and Custom." *Athenaeum* 2271 (1871): 558–59.

Anonymous. *Report of the Meeting of the British Association for the Advancement of Science.* London: John Murray, 1867.

Anonymous. "[Review of Darwin, *Expression of Emotions*]." *Times*, 13 December 1872, 4.

Anonymous. "Tylor's Primitive Culture." *North American Review* 114 (1872): 227–31.

Arnold, David. "Hodgson, Hooker and the Himalayan Frontier." In *The Origins of Himalayan Studies: Brian Houghton Hodgson in Nepal and Darjeeling, 1820–1858*, edited by David M. Waterhouse, 189–205. Abingdon: Routledge, 2004.

Arnold, David. *Science, Technology and Medicine in Colonial India.* Cambridge: Cambridge University Press, 2000.

Atkinson, Juliette. *Victorian Biography Reconsidered: A Study of Nineteenth-Century "Hidden" Lives.* Oxford: Oxford University Press, 2010.

Augstein, Hannah Franziska. *James Cowles Prichard's Anthropology: Remaking the Science of Man in Early Nineteenth-Century Britain.* Amsterdam: Rodopi, 1999.

Bachman, John. *The Doctrine of the Unity of the Human Race Examined on the Principles of Science.* Charleston, SC: Canning, 1850.

Bakhtin, M. M. *Speech Genres and Other Late Essays.* Edited by Caryl Emerson. Austin: University of Texas Press, 1986.

Ballantyne, Tony. *Orientalism and Race: Aryanism in the British Empire.* Basingstoke: Palgrave Macmillan, 2002.

Bannister, Roger. *Social Darwinism, Science and Myth in Anglo-American Social Thought.* Philadelphia: Temple University Press, 1979.

Barany, Michael. "Savage Numbers and the Evolution of Civilization in Victorian Prehistory." *British Journal for the History of Science* 47, no. 2 (2014): 239–55.

Barrett, Paul. *Metaphysics, Materialism and the Evolution of Mind: Early Writings of Charles Darwin.* Chicago: University of Chicago Press, 1980.

Barrett, Paul, Peter Gautrey, Sandra Herbert, David Kohn, and Sydney Smith, eds. *Charles Darwin's Notebooks, 1836–1844.* Cambridge: Cambridge University Press, 1987.

Barrow, Mark. *Nature's Ghosts: Confronting Extinction from the Age of Jefferson to the Age of Ecology.* Chicago: University of Chicago Press, 2009.

Basalla, George. "The Spread of Western Science." *Science* 156 (1967): 611–22.

Bayly, C. A. *Empire and Information: Intelligence Gathering and Social Communication in India, 1780–1870.* Cambridge: Cambridge University Press, 1996.

Bayly, C. A. *Recovering Liberties: Indian Thought in the Age of Liberalism and Empire.* Cambridge: Cambridge University Press, 2012.

Bayly, Susan. *Caste, Society and Politics in India from the Eighteenth Century to the Modern Age.* Cambridge: Cambridge University Press, 1999.

Beer, Gillian. *Forging the Missing Link: Interdisciplinary Stories: Inaugural Lecture, Delivered 18 November 1991.* Cambridge: Cambridge University Press Archive, 1992.

Bennett, Tony. *Pasts beyond Memory: Evolution, Museums, Colonialism.* Abingdon: Routledge, 2004.

Benton, Ted. *Alfred Russel Wallace: Explorer, Evolutionist, Public Intellectual.* Manchester: Siri Scientific Press, 2013.

Berezowsky, Sherrin. "Statistical Criticism and the Eminent Man in Francis Galton's *Hereditary Genius.*" *Victorian Literature and Culture* 43 (2015): 821–39.

Berry, Andrew, ed. *Infinite Tropics: An Alfred Russel Wallace Anthology.* London: Verso, 2002.

Bevir, Mark. "Introduction." In *Historicism and the Human Sciences in Victorian Britain*, edited by Mark Bevir, 1–20. Cambridge: Cambridge University Press, 2017.

Bieder, Robert E. *Science Encounters the Indian, 1820–1880: The Early Years of American Ethnology.* Norman: University of Oklahoma Press, 1986.

Bleichmar, Daniela. *Visible Empire: Botanical Expeditions and Visual Culture in the Hispanic Enlightenment.* Chicago: University of Chicago Press, 2012.

Blumenbach, Johann Friedrich. "On the Natural Variety of Mankind." In *The Idea of Race*, edited by Robert Bernasconi and Tommy L. Lott, 27–37. Indianapolis: Hackett, 2000.

Bollaert, William. *Antiquarian, Ethnological and Other Researches in New Granada, Equador, Peru and Chile, with Observations on the Pre-Incarial, Incarial, and Other Monuments of Peruvian Nations.* London: Trübner, 1860.

Bollaert, William. "Contributions to an Introduction to the Anthropology of the New World." *Memoirs Read before the Anthropological Society of London* 2 (1866): 92–152.

Bollaert, William. "Introduction to the Paleography of America; Or, Observations on

Ancient Picture and Figurative Writing in the New World." *Memoirs Read Before the Anthropological Society of London* 1 (1864): 169–94.
Bollaert, William. "Observations on the Past and Present Populations of the New World." *Memoirs Read before the Anthropological Society of London* 1 (1864): 72–119.
Bollaert, William. "Some Accounts of the Astronomy of the Red Man of the New World." *Memoirs Read before the Anthropological Society of London* 1 (1864): 210–80.
Bollaert, William. *Texas*. Edited by Eugene Hollon and Ruth Butler. Norman: University of Oklahoma Press, 1989.
Bowler, Peter. *Darwin Deleted: Imagining a World without Darwin*. Chicago: University of Chicago Press, 2013.
Bowler, Peter. *The Eclipse of Darwinism: Anti-Darwinian Evolution Theories in the Decades around 1900*. Baltimore: Johns Hopkins University Press, 1983.
Bowler, Peter. *The Invention of Progress: The Victorians and the Past*. Oxford: Basil Blackwell, 1989.
Bowler, Peter. *The Non-Darwinian Revolution: Reinterpreting a Historical Myth*. Baltimore: Johns Hopkins University Press, 1988.
Bowler, Peter. *Theories of Human Evolution: A Century of Debate, 1844–1944*. Baltimore: Johns Hopkins University Press, 1986.
Bravo, Michael. *The Accuracy of Ethnoscience: A Study of Inuit Cartography and Cross-Cultural Commensurability*. Manchester: Manchester Papers in Social Anthropology, 1996.
Bravo, Michael. "Ethnological Encounters." In *Cultures of Natural History*, edited by Nicholas Jardine, James Secord, and Emma Spary, 338–57. Cambridge: Cambridge University Press, 1996.
Brooke, John Hedley. *Science and Religion: Some Historical Perspectives*. Cambridge: Cambridge University Press, 1991.
Brooke, John Hedley, and Geoffrey Cantor. *Reconstructing Nature: The Engagement of Science and Religion*. Oxford: Oxford University Press, 1998.
Brooke, John Hedley, and Geoffrey Cantor. "Whose Science? Whose Religion?" In *Reconstructing Nature: The Engagement of Science and Religion*, edited by John Hedley Brooke and Geoffrey Cantor, 43–72. Oxford: Oxford University Press, 1998.
Brooks, John Langdon. *Just before the Origin: Alfred Russel Wallace's Theory of Evolution*. New York: Columbia University Press, 1984.
Brotman, Charles. "Alfred Wallace and the Anthropology of Sound in Victorian Culture." *Endeavour* 25 (2001): 144–47.

Brown, B. Ricardo. *Until Darwin, Science, Human Variety and the Origins of Race.* London: Pickering and Chatto, 2010.

Brown, Cheever Mackenzie. *Hindu Perspectives on Evolution: Darwin, Dharma, and Design.* Abingdon: Routledge, 2012.

Browne, Janet. "Biogeography and Empire." In *Cultures of Natural History*, edited by Nicholas Jardine, James Secord, and Emma Spary, 305–21. Cambridge: Cambridge University Press, 1996.

Browne, Janet. *Charles Darwin: The Power of Place.* Princeton, NJ: Princeton University Press, 2002.

Browne, Janet. *Charles Darwin: Voyaging.* London: Pimlico, 1995.

Browne, Janet. "Darwin and the Expression of the Emotions." In *The Darwinian Heritage*, edited by David Kohn, 307–26. Princeton, NJ: Princeton University Press, 1985.

Browne, Janet. "A Science of Empire: British Biogeography before Darwin." *Revue d'Histoire des Sciences* 45 (1992): 453–75.

Buckland, William. *Reliquiae Diluvianae, or Observations on the Organic Remains Contained in Caves, Fissures and Diluvial Gravel.* London: John Murray, 1824.

Buckle, Henry Thomas. *History of Civilisation in England.* 2 vols. London: J. W. Parker, 1857.

Bulmer, Michael. *Francis Galton: Pioneer of Heredity and Biometry.* Baltimore: Johns Hopkins University Press, 2003.

Burchfield, J. D. *Lord Kelvin and the Age of the Earth.* Chicago: University of Chicago Press, 1990.

Burkhardt, Frederick, et al., eds. *The Correspondence of Charles Darwin.* 30 vols. Cambridge: Cambridge University Press, 1985– .

Burrow, John W. "Evolution and Anthropology in the 1860s: The Anthropological Society of London, 1863–1871." *Victorian Studies* 7 (1963): 137–49.

Burrow, John W. *Evolution and Society: A Study in Victorian Social Theory.* Cambridge: Cambridge University Press, 1966.

Burrow, John W. "Images of Time: From Carlylean Vulcanism to Sedimentary Gradualism." In *History, Religion and Culture: British Intellectual History, 1750–1950*, edited by Stefan Collini, Richard Whatmore, Brian Young et al., 198–223. Cambridge: Cambridge University Press, 2000.

Busch, Justin E. A. *The Utopian Vision of H. G. Wells.* Jefferson, NC: McFarland, 2009.

Butler, Samuel. *Evolution, Old and New.* London: Hardwick and Bogue, 1879.

Butler, Samuel. *Life and Habit*. London: Trübner, 1878.

Calderwood, Henry. "Primitive Culture: Researches into the Development of Mythology, Philosophy, Religion, Art, and Custom." *Contemporary Review* 19 (1871): 210–22.

Caldwell, Robert. *A Comparative Grammar of the Dravidian or South-Indian Family of Languages*. London: Harrison, 1856.

Callaway, Henry. *The Religious System of the Amazulu: Izinyanga Zokubula, or, Divination, as Existing among the Amazulu, in their own Words*. London: Trübner, 1868.

Campbell, A. D. *A Grammar of the Teloogoo Language, commonly termed the Gentoo, peculiar to the Hindoos inhabiting the north eastern provinces of the Indian Peninsula*. Madras: College Press, 1816.

Cannon, Garland. *The Life and Mind of Oriental Jones: Sir William Jones, the Father of Modern Linguistics*. Cambridge: Cambridge University Press, 1990.

Cantor, Geoffrey. *Quakers, Jews and Science: Religious Responses to Modernity and the Sciences in Britain, 1650–1900*. Oxford: Oxford University Press, 2005.

Cantor, Geoffrey, Gowan Dawson, Graeme Gooday, Richard Noakes, Sally Shuttleworth, and Jonathan Topham, eds. *Science in the Nineteenth-Century Periodical: Reading the Magazine of Nature*. Cambridge: Cambridge University Press, 2004.

Cantor, Geoffrey, and Sally Shuttleworth, eds. *Science Serialized: Representations of the Sciences in Nineteenth-Century Periodicals*. Cambridge, MA: MIT Press, 2004.

Carey, Daniel. "Compiling Nature's History: Travellers and Travel Narratives in the Early Royal Society." *Annals of Science* 54 (1997): 269–92.

[Carlyle, Thomas]. *Critical and Miscellaneous Essays*. London: Chapman and Hall, 1888.

Carlyle, Thomas. "Occasional Discourse on the Negro Question," In *Critical and Miscellaneous Essays*, edited by Henry Duff Traill, 4:348–83. London: Chapman and Hall, 1899.

Carlyle, Thomas. *Past and Present*. London: Chapman and Hall, 1843.

Carlyle, Thomas. *Reminiscences of My Irish Journey in 1849*. London: Gilbert & Rivington, 1882.

[Carlyle, Thomas]. "Thoughts on History." *Fraser's Magazine* 2 (1830): 413–18.

Caso, Juan Manuel Rodríguez. "Anthropology in Transition: A Study of the Sciences of Man at the British Association for the Advancement of Science, 1866–1870." PhD diss., University of Leeds, 2014.

Caudill, Edward. "The Bishop-Eaters: The Publicity Campaign for Darwin and *On the Origin of Species*." *Journal of the History of Ideas* 55 (1994): 441–60.

Chakrabarti, Pratik. *Western Science in Modern India: Metropolitan Methods, Colonial Practices.* Delhi: Permanent Black, 2004.

Chambers, Robert. *The Vestiges of the Natural History of Creation.* London: John Churchill, 1844.

Chatterjee, Bankim Chandra. "Mill, Darwin, and Hinduism." In *Sociological Essays: Utilitarianism and Positivism in Bengal,* edited by S. N. Mukerjee and M. Maddern, 60–70. Calcutta: Rddhi, 1986.

Chatterjee, Partha. *The Nation and Its Fragments: Colonial and Postcolonial Histories.* Princeton, NJ: Princeton University Press, 1993.

Chatterjee, Partha. *Nationalist Thought and the Colonial World: A Derivative Discourse?* London: Zed Books, 1986.

Chesterton, G. K. *Eugenics and Other Evils.* London: Cassell, 1922.

Chidester, David. "Darwin's Dogs: Animals, Animism, and the Problem of Religion." *Soundings: An Interdisciplinary Journal* 92 (2009): 51–75.

Cleall, Esme. *Missionary Discourses of Difference: Negotiating in the British Empire, 1840–1900.* Basingstoke: Palgrave Macmillan, 2012.

Cohn, Bernard. *Colonialism and Its Forms of Knowledge: The British in India.* Princeton, NJ: Princeton University Press, 1996.

Comte, Auguste. *Cours de Philosophie Positive.* 6 vols. Paris: Bachelier, 1830–1842.

Cornell, John. "From Creation to Evolution: Sir William Dawson and the Idea of Design in the Nineteenth Century." *Journal of the History of Biology* 16, no. 1 (1983): 137–70.

Costa, James T. "Engaging with Lyell: Alfred Russel Wallace's Sarawak Law and Ternate Papers as Reactions to Charles Lyell's Principles of Geology." *Theory in Biosciences* 132, no. 4 (2013): 225–37.

Costa, James T. *Wallace, Darwin, and the Origin of Species.* Cambridge, MA: Harvard University Press, 2014.

Cowan, Ruth Schwartz. "Galton, Sir Francis (1822–1911)." *Oxford Dictionary of National Biography.* Oxford: Oxford University Press, 2005.

Dalton, Edward Tuite. *Descriptive Ethnology of Bengal.* Calcutta: Office of the Superintendent of Government Printing, 1872.

Darwin, Charles. *The Autobiography of Charles Darwin, 1809–1882, with Original Omissions Restored.* Edited by Nora Barlow. London: Collins, 1958.

Darwin, Charles. *The Descent of Man, and Selection in Relation to Sex.* 2 vols. London: John Murray, 1871.

Darwin, Charles. *The Expression of the Emotions in Man and Animals.* London: John Murray, 1872.

Darwin, Charles. *Journal of Researches into the Natural History and Geology of the Countries Visited During the Voyage of H.M.S.* Beagle *Round the World.* 2nd ed. London: John Murray, 1845.

Darwin, Charles. *On the Origin of Species by Means of Natural Selection.* London: John Murray, 1859.

Darwin, Charles. *The Variation of Animals and Plants under Domestication.* 2 vols. London: John Murray, 1868.

Darwin Correspondence Database. "Hooker, Joseph D. to Darwin, Charles. 14 May 1864." http://www.darwinproject.ac.uk/entry-4494.

Dawson, Gowan. *Darwin, Literature and Victorian Respectability.* Cambridge: Cambridge University Press, 2007.

Dawson, Gowan, and Bernard Lightman, eds. *Victorian Scientific Naturalism: Community, Identity, and Continuity.* Chicago: University of Chicago Press, 2014.

Dawson, John William. *Archaia: Or, Studies of the Cosmogony and Natural History of the Hebrew Scriptures.* Montreal: B. Dawson & Son, 1860.

Dawson, John William. "Review of Darwin, *On the Origin of Species by Means of Natural Selection.*" *Canadian Naturalist and Geologist* 5 (1860): 100–120.

Dawson, John William. *The Story of the Earth and Man.* New York: Harper and Brothers, 1873.

Dawkins, William Boyd. "The British Lion." *Popular Science* (November 1882): 72–81.

Dawkins, William Boyd. *Cave Hunting: Researches on the Evidence of Caves Respecting the Early Inhabitants of Europe.* London: Macmillan, 1874.

Dawkins, William Boyd. *Early Man in Britain and His Place in the Tertiary Period.* London: Macmillan, 1880.

Desmond, Adrian. *Huxley: From Devil's Disciple to Evolution's High Priest.* London: Penguin Books, 1998.

Desmond, Adrian, and James Moore. *Darwin: The Life of a Tormented Evolutionist.* New York: Warner, 1991.

Desmond, Adrian, and James Moore. *Darwin's Sacred Cause: How a Hatred of Slavery Shaped Darwin's Views on Human Evolution.* Boston: Houghton Mifflin Harcourt, 2009.

Desmond, Adrian, and James Moore. *Darwin's Sacred Cause: Race, Slavery and the Quest for Human Origins.* London: Penguin Books, 2009.

Díaz-Andreu, Margarita. *A World History of Nineteenth-Century Archaeology: Nationalism, Colonialism, and the Past.* Oxford: Oxford University Press, 2007.

DiGregorio, Mario. *T. H. Huxley's Place in Natural Science.* New Haven: Yale University Press, 1984.

Dixon, Mike, and Gregory Radick. *Darwin in Ilkley.* Stroud: History Press, 2009.

Dodson, Michael S. *Orientalism, Empire, and National Culture: India, 1770–1880.* Basingstoke: Palgrave Macmillan, 2007.

Douglas, Kirsty. *Pictures of Time Beneath: Science, Heritage and the Uses of the Deep Past.* Clayton: Csiro Publishing, 2010.

Draper, John William. *History of the Conflict between Religion and Science.* New York: D. Appleton, 1874.

Drayton, Richard. *Nature's Government: Science, Imperial Britain, and the "Improvement" of the World.* Telangana: Orient Longman, 2005.

Dugger, Julie. "Black Ireland's Race: Thomas Carlyle and the Young Ireland Movement." *Victorian Studies* 48 (2006): 461–85.

Duns, John. "Archaia; Or, Studies of the Cosmogony and Natural History of the Hebrew Scriptures." *North British Review* 32, no. 64 (May 1860): 545–48.

Edney, Matthew. *Mapping an Empire: The Geographical Construction of British India, 1765–1843.* Chicago: University of Chicago Press, 1997.

Eiseley, Loren C. *Darwin's Century: Evolution and the Men Who Discovered It.* New York: Doubleday, 1958.

Ekman, Paul. "Introduction (and Commentaries)." In *The Expression of the Emotions in Man and Animals*, 3rd ed., by Charles Darwin, xxi–xxxvi. London: HarperCollins, 1998.

Ellegård, Alvar. *Darwin and the General Reader: The Reception of Darwin's Theory of Evolution in the British Periodical Press, 1859–1872.* Chicago: University of Chicago Press, 1990.

Ellen, Roy. "The Place of the Neolithic Controversy in the Anthropology of Alfred Russel Wallace." *Linnean* 27 (2011): 22–33.

Ellis, William. *Polynesian Researches during a Residence of Nearly Six Years in the South Sea Islands.* 3 vols. London: Fisher, Son and Jackson, 1829–1831.

Elshakry, Marwa. *Reading Darwin in Arabic.* Chicago: University of Chicago Press, 2013.

Elwick, James. *Styles of Reasoning in the British Life Sciences: Shared Assumptions, 1820–58.* London: Pickering and Chatto, 2007.

Fabian, Anne. *The Skull Collectors: Race, Science and America's Unburied Dead.* Chicago: University of Chicago Press, 2010.

Fagan, Melinda B. "Wallace, Darwin, and the Practice of Natural History." *Journal of the History of Biology* 40, no. 4 (2007): 601–35.

Fara, Patricia. "[Review of Desmond and Moore, *Darwin's Sacred Cause*]." *Journal of Historical Geography* 35 (2009): 614–16.

Farrall, Lyndsay Andrew. *The Origins and Growth of the English Eugenics Movement, 1865–1925.* London: Garland, 1985.

Fetherling, George. *The Rise of the Canadian Newspaper.* Oxford: Oxford University Press, 1990.

Fichman, Martin. "Biology and Politics: Defining the Boundaries." In *Victorian Science in Context,* edited by Bernard Lightman, 94–118. Chicago: University of Chicago Press, 1997.

Fichman, Martin. *An Elusive Victorian: The Evolution of Alfred Russel Wallace.* Chicago: University of Chicago Press, 2004.

Figuier, Louis. *L'homme primitif.* Paris: Librairie de L. Hachette, 1870.

Figuier, Louis. *Primitive Man.* London: Chapman and Hall, 1870.

Fleming, Patricia, Yvan Lamonde, and Fiona Black, eds. *History of the Book in Canada, 1840–1918.* Toronto: University of Toronto Press, 2005.

Forkey, Neil S. *Canadians and the Natural Environment to the Twenty-First Century.* Toronto: University of Toronto Press, 2012.

Forrest, D. W. *Francis Galton: The Life and Work of a Victorian Genius.* London: Taplinger, 1974.

Francis, Mark, and Michael W. Taylor, eds. *Herbert Spencer: Legacies.* Abingdon: Routledge, 2015.

Franco, Jean. "Un viaje poco romántico: Viajeros británicos hacia Sudamérica, 1818–28." *Escritura, Teoría y Crítica Literaria IV,* 7 (1979): 129–42.

Freeman, Richard B., and Peter J. Gautrey. "Charles Darwin's *Queries about Expression.*" *Bulletin of the British Museum (Natural History),* Historical Series 4 (1972): 205–19.

Freeman Richard B., and Peter J. Gautrey. "Charles Darwin's *Queries about Expression.*" *Journal of the Society for the Bibliography of Natural History* 7 (1975): 259–63.

Fridlund, Alan. "Darwin's Anti-Darwinism in *The Expression of the Emotions in Man and Animals.*" In *International Review of Studies on Emotion,* edited by K. T. Strongman, 2:117–37. Chichester: Wiley, 1992.

Fyfe, Aileen. *Science and Salvation: Evangelical Popular Science Publishing in Victorian Britain.* Chicago: University of Chicago Press, 2004.

Fyfe, Aileen, and Bernard Lightman, eds. *Science in the Marketplace: Nineteenth-Century Sites and Experiences.* Chicago: University of Chicago Press, 2007.

Gaenszle, Martin. "Brian Hodgson as Ethnographer and Ethnologist." In *The Origins of Himalayan Studies: Brian Houghton Hodgson in Nepal and Darjeeling, 1820–1858,* edited by David M. Waterhouse, 206–26. Abingdon: Routledge, 2004.

Gallagher, Catherine. *The Industrial Reformation of English Fiction: Social Discourses and Narrative Form, 1832–1867.* Chicago: University of Chicago Press, 1985.

Galton, Francis. "Eugenics as a Factor in Religion." In *Essays in Eugenics,* by Francis Galton. London: Eugenics Education Society, 1909.

Galton, Francis. *Finger Prints.* London: Macmillan, 1892.

Galton, Francis. *Hereditary Genius: An Inquiry into Its Laws and Consequences.* London: Macmillan, 1869.

Galton, Francis. *Hereditary Genius.* Rev. ed. London: Macmillan, 1892.

Galton, Francis. "Hereditary Talent and Character." *Macmillan's Magazine* 12 (1865): 157–66.

Galton, Francis. *Inquiries into Human Faculty and Its Development.* New York: Macmillan, 1883.

Gaskell, Elizabeth. *Mary Barton.* Edited by Jennifer Foster. Peterborough: Broadview, 2000.

Gaskell, Elizabeth. "The Well of Pen-Morfa." In *The Works of Gaskell,* edited by A. W. Ward, 2:242–66. London: John Murray, 1920.

Ghose, Aurobindo. *Evolution.* Calcutta: Arya, 1944.

Gillham, Nicholas Wright. *A Life of Sir Francis Galton: From African Exploration to the Birth of Eugenics.* Oxford: Clarendon Press, 2001.

Glendening, John. *The Evolutionary Imagination in Late-Victorian Novels: An Entangled Bank.* Aldershot: Ashgate, 2016.

Glick, Thomas F., ed. *The Comparative Reception of Darwin.* Chicago: University of Chicago Press, 1988.

Gökyiğit, Emel Aileen. "Reception of Francis Galton's 'Hereditary Genius' in the Victorian Periodical Press." *Journal of the History of Biology* 27 (1994): 215–40.

Goldman, Lazarus. *The History of Jews in New Zealand.* Wellington: Reed, 1958.

Goodall, Jane. *Performance and Evolution in the Age of Darwin: Out of the Natural Order.* Abingdon: Routledge, 2002.

Gosling, David L. "Darwin and the Hindu Tradition: 'Does What Goes around Come Around?'" *Zygon* 46 (2011): 345–69.

Gould, Stephen Jay. *Dinosaur in a Haystack: Reflections in Natural History.* London: Jonathan Cape, 1996.

Gould, Stephen Jay. *The Mismeasure of Man.* New York: W. W. Norton, 1981.

Gould, Stephen Jay. *Ontogeny and Phylogeny.* Cambridge, MA: Harvard University Press, 1977.

Gould, Stephen Jay. *Time's Arrow, Time's Cycle: Myth and Metaphor in the Discovery of Geological Time.* Cambridge, MA: Harvard University Press, 1987.

Grayson, Donald. *The Establishment of Human Antiquity.* New York: Academic Press, 1983.

Greene, Mott T. *Geology in the Nineteenth Century: Changing Views of a Changing World.* Ithaca: Cornell University Press, 1982.

Grondin, Jean. "Gadamer's Basic Understanding of Understanding." In *The Cambridge Companion to Gadamer*, edited by Robert J. Dostal, 36–51. Cambridge: Cambridge University Press, 2002.

Grote, George. *A History of Greece: From the Earliest Period to the Close Generation Contemporary with Alexander the Great.* 12 vols. London: John Murray, 1846–1856.

Gruber, Howard. *Darwin on Man: A Psychological Study of Creativity.* London: Wildwood House, 1974.

Guha, Sumit. "Lower Strata, Older Races, and Aboriginal Peoples: Racial Anthropology and Mythical History Past and Present." *Journal of Asian Studies* 57 (1998): 423–41.

Hale, Piers. *Political Descent: Malthus, Mutualism, and the Politics of Evolution in Victorian England.* Chicago: University of Chicago Press, 2014.

Hall, Catherine. *Civilizing Subjects: Metropole and Colony in the English Imagination, 1830–1867.* Chicago: University of Chicago Press, 2002.

Hall, Catherine, and Sonya Rose. "Introduction: Being at Home with the Empire." In *At Home with the Empire: Metropolitan Culture and the Imperial World*, edited by Catherine Hall and Sonya Rose, 1–31. Cambridge: Cambridge University Press, 2006.

Hamilton, Alexander. "Review of Francis Buchanan's 'A Journey from Madras.'" *Edinburgh Review* (October 1808): 82–100.

Hampton, Monte Harrell. *Storm of Words: Science, Religion, and Evolution in the Civil War Era.* Tuscaloosa: University of Alabama Press, 2014.

Harrison, Peter. "'Science' and 'Religion': Constructing the Boundaries." *Journal of Religion* 86 (2006): 81–106.

Hartley, L. P. *The Go-Between*. Harmondsworth: Penguin, 1958.

Hawkins, Mike. *Social Darwinism in European and American Thought, 1860–1945*. Cambridge: Cambridge University Press, 1997.

Hazlitt, William. *Spirit of the Age; or, Contemporary Portraits*. Plymouth: Northcote House, 1990.

Headrick, Daniel. *Power over Peoples: Technology, Environments, and Western Imperialism, 1400 to the Present*. Princeton, NJ: Princeton University Press, 2009.

Hechter, Michael. *Internal Colonialism: The Celtic Fringe in British National Development, 1536–1966*. Berkeley: University of California Press, 1975.

Henderson, Gerald M. "Alfred Russel Wallace: His Role and Influence in Nineteenth Century Evolutionary Thought." PhD diss., University of Pennsylvania, 1958.

Hertz, Friedrich Otto. *Race and Civilization*. New Jersey: Ktav, 1970.

Hesketh, Ian. "A Good Darwinian? Winwood Reade and the Making of a Late Victorian Evolutionary Epic." *Studies in History and Philosophy of Biological and Biomedical Sciences* 51 (2015): 44–52.

Hesketh, Ian. "The Recurrence of the Evolutionary Epic." *Journal of the Philosophy of History* 9, no. 2 (2015): 196–219.

Hesketh, Ian. *The Science of History in Victorian Britain*. London: Pickering and Chatto, 2011.

Hodge, M. J. S. *Before and after Darwin: Origins, Species, Cosmogonies, and Ontologies*. Aldershot: Ashgate, 2008.

Hodge, M. J. S. "Capitalist Contexts for Darwinian Theory: Land, Finance, Industry and Empire." *Journal of the History of Biology* 42, no. 3 (2009): 399–416.

Hodge, M. J. S. "Chance and Chances in Darwin's Early Theorizing and in Darwinian Theory Today." In *Chance in Evolution*, edited by Grant Ramsey and Charles H. Pence, 41–75. Chicago: University of Chicago Press, 2016.

Hodge, M. J. S. "Darwin, the Galápagos and His Changing Thoughts about Species Origins, 1835–1837." *Proceedings of the California Academy of Sciences* 61 (2010): 89–106.

Hodge, M. J. S. "The Notebook Programmes and Projects of Darwin's London Years." In *The Cambridge Companion to Darwin*, edited by Jonathan Hodge and Gregory Radick, 44–72. 2nd ed. Cambridge: Cambridge University Press, 2009.

Hodgkin, Thomas, and Richard Cull. "A Manual of Ethnological Inquiry." *Journal of the Ethnological Society of London* 3 (1854): 193–208.

Hopps, John Page. *The Future Life*. London: Simpkin, Marshall, [1884].

Hulme, Peter, and Tim Youngs. "Introduction." In *Cambridge Companion to Travel Writing*, edited by Peter Hulme and Tim Youngs, 1–16. Cambridge: Cambridge University Press, 2002.

Hunt, James. "The President's Address." *Journal of the Anthropological Society of London* 2 (1864): lxxx–xcv.

Huth, Alfred Henry. *The Life and Writings of Henry Thomas Buckle*. 2 vols. London: Sampson, Low, Marston, Searle & Rivington, 1880.

Huxley, Thomas. *Evidence as to Man's Place in Nature*. London: Williams and Norgate, 1863.

Huxley, Thomas. *Evolution and Ethics*. London: Macmillan, 1894.

Huxley, Thomas. "On the Classification and Distribution of the Alectoromorphae and Heteromorphae." *Proceedings of the Zoological Society of London* (1868): 294–319.

Ione, Amy. *Art and the Brain: Plasticity, Embodiment and the Unclosed Circle*. Leiden: Brill, 2016.

Isichei, Elizabeth. *Victorian Quakers*. Oxford: Oxford University Press, 1970.

Johnson, Robert. *British Imperialism*. Basingstoke: Palgrave Macmillan, 2003.

Johnston, Anna. *Missionary Writing and Empire, 1800–1860*. Cambridge: Cambridge University Press, 2003.

Jones, Kristine. "Nineteenth Century British Travel Accounts of Argentina." *Ethnohistory* 33 (1986): 195–211.

Jones, Thomas, ed. *Reliquiae Aquitanicae: Being Contributions to the Archaeology and Palaeontology of Périgord and the Adjoining Provinces of Southern France*. London: Williams & Norgate, 1875.

Jones, William. *The Works of Sir William Jones. In Six Volumes*, edited by Anna Maria Shipley Jones. 6 vols. London: G. G. and J. Robinson, 1799.

Kaalund, Nanna Katrine Lüders. "Oxford Serialised: Revisiting the Huxley-Wilberforce Debate through the Periodical Press." *History of Science* 52 (2014): 429–53.

Kapila, Shruti. "Race Matters: Orientalism and Religion, India and Beyond, c.1770–1880." *Modern Asian Studies* 41 (2007): 471–513.

Kapila, Shruti. "Self, Spencer and Swaraj: Nationalist Thought and Critiques of Liberalism, 1890–1920." *Modern Intellectual History* 4 (2007): 109–27.

Kenny, Robert. "From the Curse of Ham to the Curse of Nature: The Influence of Natural Selection on the Debate on Human Unity before the Publication of *The Descent of Man*." *British Journal for the History of Science* 40 (2007): 363–88.

Kevles, Daniel J. *In the Name of Eugenics: Genetics and the Uses of Human Heredity.* New York: Knopf, 1985.

Keynes, Milo, ed. *Sir Francis Galton, FRS: The Legacy of His Ideas.* Basingstoke: Palgrave Macmillan, 1993.

Kidd, Colin. *The Forging of Races: Race and Scripture in the Protestant Atlantic World, 1600–2000.* Cambridge: Cambridge University Press, 2006.

Kjærgaard, Peter C. K. "'Hurrah for the Missing Link!' A History of Apes, Ancestors and a Crucial Piece of Evidence." *Notes and Records the Royal Society Journal of the History of Science* 65 (2011): 83–98.

Knipe, Henry. *Evolution in the Past.* London: Herbert and Daniel, 1912.

Knox, Robert. *The Races of Man: A Fragment.* London: Henry Renshaw, 1850.

Koditschek, Theodore. *Liberalism, Imperialism and the Historical Imagination: Nineteenth-Century Visions of a Greater Britain.* Cambridge: Cambridge University Press, 2011.

Koditschek, Theodore. "Narrative Time and Racial/Evolutionary Time in Nineteenth-Century British Liberal Imperial History." In *Race, Nation and Empire: Making Histories, 1750 to the Present*, edited by Catherine Hall and Keith McClelland, 36–55. Manchester: Manchester University Press, 2010.

Kohn, Margaret. "Colonialism." *The Stanford Encyclopedia of Philosophy* (Spring 2014), ed. Edward N. Zalta. http://plato.stanford.edu/archives/spr2014/entries/colonialism/.

Kottler, Malcolm Jay. "Alfred Russel Wallace, the Origin of Man, and Spiritualism." *Isis* 65, no. 2 (1974): 145–92.

Kuklick, Henrika. *The Savage Within: The Social History of British Anthropology, 1885–1945.* Cambridge: Cambridge University Press, 1991.

Kuklick, Henrika. "The Theory of Evolution and Cultural Anthropology." In *The Theory of Evolution and Its Impact*, edited by Aldo Fasolo, 83–102. New York: Springer Verlag, 2012.

Kumar, Deepak. "Reason, Science and Religion: Gleanings from the Colonial Past." *Current Science* 99 (2010): 671–78.

Lafont-Couturier, Hélène, and Katia Bosch, eds. *Vénus et Caïn: Figures de la Préhistoire, 1830–1930.* Bordeaux: Réunion des musées nationaux, 2003.

Landau, Misia. *Narratives of Evolution.* New Haven: Yale University Press, 1991.

Lang, Andrew. "Edward Burnett Tylor." In *Anthropological Essays Presented to Edward Tylor in Honour of his 75th Birthday*, edited by Henry Balfour, 1–17. Oxford: Clarendon Press, 1907.

Lankester, E. Ray. *Degeneration: A Chapter in Darwinism.* London: Macmillan, 1880.

Lankester, E. Ray. *Kingdom of Man.* New York: Henry Holt, 1907.

Lartet, Eduard. "Nouvelles Recherches sur la Coexistence de l'Homme avec des Grands Mammifères Fossiles Réputés Caractéristiques de la Dernière Période Géologique." *Annales des sciences naturelles, quatrième série: Zoologie* 15 (1861): 176–253.

Leopold, Joan. *Culture in Comparative and Evolutionary Perspective: E. B. Tylor and the Making of Primitive Culture.* Berlin: Dietrich Reimer Verlag, 1980.

Lewis, George C. "The Astronomy of the Ancients." *Ethnological Journal: A Monthly Record of Ethnological Research and Criticism* (1865): 211–21.

Hong-Jin Liu. "Data and the Development of Research Methods in the Science of Human Emotional Expression from Darwin to Klineberg." PhD diss., University of Leeds, 2016.

Lightman, Bernard. "Darwin and the Popularization of Evolution." *Notes and Records of the Royal Society of London* 64 (2010): 5–24.

Lightman, Bernard. "Does the History of Science and Religion Change depending on the Narrator? Some Atheist and Agnostic Perspectives." *Science and Christian Beliefs* 24 (2012): 149–68.

Lightman, Bernard, ed. *Global Spencerism: The Communication and Appropriation of a British Evolutionist.* Leiden: Brill, 2015.

Lightman, Bernard. "The 'History' of Victorian Scientific Naturalism: Huxley, Spencer, and the 'End' of Natural History." *Studies in History and Philosophy of Biological and Biomedical Sciences* 58 (2016): 17–23.

Lightman, Bernard. *The Origins of Agnosticism: Victorian Unbelief and the Limits of Knowledge.* Baltimore: Johns Hopkins University Press, 1987.

Lightman, Bernard. "Scientists as Materialists in the Periodical Press: Tyndall's Belfast Address.," In *Science Serialized: Representations of the Sciences in Nineteenth-Century Periodicals*, edited by Geoffrey Cantor and Sally Shuttleworth, 199–237. Cambridge: MIT Press, 2004.

Lightman, Bernard. *Victorian Popularizers of Science: Designing Nature for New Audiences.* Chicago: University of Chicago Press, 2007.

Lightman, Bernard. *Victorian Science in Context.* Chicago: University of Chicago Press, 1997.

Lightman, Bernard. "Victorian Sciences and Religion: Discordant Harmonies." *Osiris* 16 (2001): 343–66.

Lindberg, David C., and Ronald L. Numbers, eds. *When Science and Christianity Meet.* Chicago: University of Chicago Press, 2003.

Livingstone, David. *Adam's Ancestors: Race, Religion, and the Politics of Human Origins.* Baltimore: Johns Hopkins University Press, 2008.

Livingstone, David. *Dealing with Darwin: Place, Politics, and Rhetoric in Religious Engagements with Evolution.* Baltimore: Johns Hopkins University Press, 2014.

Livingstone, David, and Charles W. J. Withers. *Geographies of Nineteenth-Century Science.* Chicago: University of Chicago Press, 2011.

Lorimer, Douglas. *Science, Race Relations and Resistance: Britain, 1870–1914.* Manchester: Manchester University Press, 2013.

Lowell, Percivall. *Mars and Its Canals.* New York: Macmillan, 1906.

Lowrey, Kathleen Bolling. "Alfred Russel Wallace as Ancestor Figure: Reflections on Anthropological Lineage after the Darwin Bicentennial." *Anthropology Today* 26 (2010): 18–21.

Lubbock, John. *Pre-Historic Times: As Illustrated by Ancient Remains, and the Manners and Customs of Modern Savages.* London: Williams and Norgate, 1865.

Lyell, Charles. *The Geological Evidences of the Antiquity of Man.* London: John Murray, 1863.

[Macaulay, Thomas Babington]. "Ecclesiastical and Political History of the Popes of Rome during the Sixteenth and Seventeenth Centuries." *Edinburgh Review* 72 (1840): 227–58.

[Macaulay, Thomas Babington]. "The Romance of History." *Edinburgh Review* 47 (1828): 331–67.

MacLeod, Roy. "Introduction: Nature and Empire: Science and the Colonial Enterprise." *Osiris* 15 (2000): 1–13.

Macpherson, James. *The Poems of Ossian.* London: Cadell and Davies, 1807.

Maine, Henry James Sumner. *Ancient Law: Its Connection with the Early History of Society and its Relations to Modern Ideas.* London: John Murray, 1861.

Manias, Chris. "The Problematic Construction of 'Palaeolithic Man': The Old Stone Age and the Difficulties of the Comparative Method, 1859–1914." *Studies in History and Philosophy of Biological and Biomedical Sciences* 51 (2015): 32–43.

Manias, Chris. *Race, Science, and the Nation: Reconstructing the Ancient Past in Britain, France and Germany.* London: Routledge, 2013.

Mantena, Karuna. *Alibis of Empires: Henry Maine and the Ends of Liberal Imperialism.* Princeton, NJ: Princeton University Press, 2010.

Mantena, Rama Sundari. *The Origins of Modern Historiography in India: Antiquarianism and Philology, 1780–1880*. New York: Palgrave and Macmillan, 2012.

Marchant, James, ed. *Alfred Russel Wallace Letters and Reminiscences*. 2 vols. London: Cassell, 1916.

Marett, Robert Ranulph. *Tylor.* London: Chapman and Hall, 1936.

Marx, Karl. *Capital.* Vol. 1. Moscow: Progress Moscow, 1954

Marx, Karl. *A Contribution to the Critique of Political Economy.* New York: International Press, 1970.

Marx, Karl, and Friedrich Engels. *The Communist Manifesto.* New York: International Press, 1948.

Matthew, H. C. G. "Walpole, Sir Spencer (1839–1907)." *Oxford Dictionary of National Biography*. Oxford: Oxford University Press, 2004. http://www.oxforddnb.com/view/article/36712/.

Mayr, Ernst. "Wallace's Line in the Light of Recent Zoogeographic Studies." *Quarterly Review of Biology* 19 (1944): 1–14.

McClintock, Anne. *Imperial Leather: Race, Gender, and Sexuality in the Colonial Contest.* Abingdon: Routledge, 1995.

McKinney, Henry Lewis. *Wallace and Natural Selection.* New Haven: Yale University Press, 1972.

McKinney, Henry Lewis. "Wallace's Earliest Observations on Evolution, 28 December 1845." *Isis* 60 (1969): 370–73.

McLennan, John Ferguson. *Primitive Marriage: An Inquiry into the Origin of the Form of Capture in Marriage Ceremonies.* Edinburgh: Adam and Charles Black, 1865.

McNabb, John. *Dissent with Modification: Human Origins, Palaeolithic Archaeology and Evolutionary Anthropology in Britain, 1859–1901.* Oxford: Archaeopress, 2012.

McPhee, John. *Basin and Range.* New York: Farrar, Straus, Giroux, 1981.

Meek, Ronald L. *Social Science and the Ignoble Savage.* Cambridge: Cambridge University Press, 1976.

[Merivale, Henry]. "Galton on Hereditary Genius." *Edinburgh Review* 132 (1870): 52–64.

Metcalfe, Thomas. *Ideologies of the Raj.* Cambridge: Cambridge University Press, 1994.

Midgley, Mary. *Animals and Why They Matter.* Atlanta: University of Georgia Press, 1984.

Miller, David Phillip, and Peter Hanns Reill, eds. *Visions of Empire: Voyages, Botany, and Representations of Nature.* Cambridge: Cambridge University Press, 1996.

Miller, Hugh. *Hugh Miller's Memoir: From Stonemason to Geologist.* Edinburgh: Edinburgh University Press, 1995.

Mivart, St. George Jackson. *On the Genesis of Species.* London: Macmillan, 1871.

Montgomery, William. "Charles Darwin's Thought on Expressive Mechanisms in Evolution." In *The Development of Expressive Behavior: Biology-Environment Interactions,* edited by Gail Zivin, 27–50. London: Academic Press, 1985.

Moretti, Franco. *Atlas of the European Novel, 1800–1900.* New York: Verso, 1999.

Morrell, Jack, and Arnold Thackray. *Gentlemen of Science: Early Years of the British Association for the Advancement of Science.* Oxford: Oxford University Press, 1981.

Moore, James. *The Post-Darwinian Controversies: A Study of the Protestant Struggle to Come to Terms with Darwin in Great Britain and America, 1879–1900.* Cambridge: Cambridge University Press, 1979.

Moore, James. "Wallace's Malthusian Moment: The Common Context Revisited." In *Victorian Science in Context,* edited by Bernard Lightman, 290–311. Chicago: University of Chicago Press, 1997.

Murray, Tim. "Illustrating 'Savagery': Sir John Lubbock and Ernest Griset." *Antiquity* 83 (2008): 488–99.

Myers, Frederic W. H. "Automatic Writing." *Proceedings of the Society for Psychical Research* 3 (1885): 1–63.

Myers, Frederic W. H. "Science and a Future Life." *Nineteenth Century* (April 1891): 628–47.

Myers, Frederic W. H. *Science and a Future Life with Other Essays.* London: Macmillan, 1893.

Myers, Greg. "Nineteenth-Century Popularizations of Thermodynamics and the Rhetoric of Social Prophecy." *Victorian Studies* 29, no. 1 (1985): 36–66.

Neufville, John Bryan. "On the Geography and Population of Asam." *Asiatick Researches* 16 (1828): 331–52.

Numbers, Ronald L. *The Creationists.* Berkeley: University of California Press, 1993.

Numbers, Ronald L., ed. *Galileo Goes to Jail and Other Myths about Science and Religion.* Cambridge, MA: Harvard University Press, 2009.

Numbers, Ronald L., ed. *When Science and Christianity Meet.* Chicago: University of Chicago Press, 2003.

Numbers, Ronald L., and John Stenhouse, eds. *Disseminating Darwinism: The Role of Place, Race, Religion, and Gender.* Cambridge: Cambridge University Press, 1999.

O'Brien, Charles F. *Sir William Dawson, a Life in Science and Religion.* Philadelphia: American Philosophical Society, 1971.

O'Connor, Ralph. *The Earth on Show: Fossils and the Poetics of Popular Science, 1802–1856.* Chicago: University of Chicago Press, 2007.

Oldroyd, David. *Thinking about the Earth: A History of Ideas in Geology.* Cambridge, MA: Harvard University Press, 1996.

Oppenheim, Janet. *The Other World: Spiritualism and Psychical Research in England, 1850–1914.* Cambridge: Cambridge University Press, 1985.

O'Reilly, Clare. "Darwin, Dogs and the Tree of Life: A Revisionist Account of the Role of Cross-Breeding and Hybridism in Darwin's Evolutionary Theorising." MA diss., University of Leeds, 2014.

Owen, Richard. *Description of the Skeleton of an Extinct Gigantic Sloth,* Mylodon Robustus. London: John Van Voorst, 1842.

Owen, Richard. *A History of British Fossil Mammals and Birds.* London: John Van Voorst, 1846.

Owen, Richard. *Palaeontology, or, a Systematic Summary of Extinct Animals and Their Geological Relations.* Edinburgh: Adam and Charles Black, 1860.

Pagden, Anthony. *The Fall of Natural Man: The American Indian and the Origins of Comparative Ethnology.* Cambridge: Cambridge University Press, 1982.

Palacio-Pérez, Eduardo. "Salomon Reinach and the Religious Interpretation of Palaeolithic Art." *Antiquity* 84 (2010): 853–63.

Parkyn, Ernest A. *An Introduction to the Study of Prehistoric Art.* London: Longmans, Green, 1915.

Paul, Diane B. *The Politics of Heredity: Essays on Eugenics, Biomedicine, and the Nature-Nurture Debate.* Albany: State University of New York Press, 1998.

Pearson, Karl. *The Life, Letters and Labours of Francis Galton.* 3 vols. Cambridge: Cambridge University Press, 1914.

Peel, J. D. Y. *Herbert Spencer: The Evolution of a Sociologist.* New York: Basic Books, 1971.

Penman, Leigh T. I. "The History of the Word *Paranormal.*" *Notes and Queries* 260, no. 1 (March 2015): 31–34.

Pettit, Paul, and Mark White. "Cave Men: Stone Tools, Victorian Science, and the 'Primitive Mind' of Deep Time." *Notes and Records: The Royal Society Journal* 65 (2010): 25–42.

Pick, Daniel. *Faces of Degeneration: A European Disorder, c.1848–c.1918.* Cambridge: Cambridge University Press, 1989.

Pimentel, Juan. "Across Nations and Ages: The Creole Collector and the Many Lives of the Megatherium." In *The Brokered World: Go-Betweens and Global Intelligence, 1770–1820*, edited by Simon Schaffer, Lissa Roberts, Kapil Raj, and James Delbourgo, 321–54. Sagamore Beach: Science History Publications, 2009.

Pinker, Steven. "Still Stimulating after All These Years." Review of Ekman's edition of the *Expression*, 3rd ed., edited by Paul Ekman. *Science* 281 (24 July 1998): 522–23.

Podgorny, Irina. "Fossil Dealers: The Practices of Comparative Anatomy and British Diplomacy in Latin America, 1820–1840." *British Journal for the History of Science* 46 (2013): 647–74.

Polack, Joel Samuel. *Manners and Customs of the New Zealanders; with Notes Corroborative of their Habits, Usages Etc.* 2 vols. London: James Madden, 1840.

Poskett, James. "Metropole of the Mind: Phrenology and the Making of a Global Science, 1815–1860." PhD diss., University of Cambridge, 2015.

Postans, T. "On the Biluchi Tribes Inhabiting Sindh in the Lower Valley of the Indus and Cutchi." *Journal of the Ethnological Society of London* 1 (1848): 103–26.

Prakash, Gyan. *Another Reason: Science and the Imagination of Modern India.* Princeton, NJ: Princeton University Press, 1999.

Pratt, Mary Louise. *Imperial Eyes: Travel Writing and Transculturation.* Abingdon: Routledge, 1992.

Price, Michael. "Facial Expressions—Including Fear—May Not Be as Universal as We Thought." *Science* (17 October 2016). http://www.sciencemag.org/news/2016/10/facial-expressions-including-fear-may-not-be-universal-we-thought.

Prichard, James Cowles. *Researches into the Physical History of Man.* London: John and Arthur Arch, 1813.

Prichard, James Cowles. *Researches into the Physical History of Mankind.* 3rd ed. 5 vols. London: Sherwood, Gilbert, and Piper, 1836–1847.

Radick, Gregory. "Animal Agency in the Age of the Modern Synthesis: W. H. Thorpe's Example." In *Animal Agents: The Non-human in the History of Science*, edited by Amanda Rees. British Journal for the History of Science Themes 2. Cambridge: Cambridge University Press, 2017.

Radick, Gregory. "Darwin and Humans." In *The Cambridge Encyclopedia of Darwin and Evolutionary Thought*, edited by Michael Ruse, 173–81. Cambridge: Cambridge University Press, 2013.

Radick, Gregory. "Darwin on Language and Selection." *Selection* 3 (2002): 7–16.

Radick, Gregory. "Darwin's Puzzling *Expression*." *Comptes Rendus Biologies* 333 (2010): 181–87.

Radick, Gregory. "Did Darwin Change His Mind about the Fuegians?" *Endeavour* 34 (2010): 50–54.

Radick, Gregory. *Edinburgh and Darwin's* Expression of the Emotions. Pamphlet in the Dialogues with Darwin Series. Edinburgh: Institute for Advanced Studies in the Humanities, 2009.

Radick, Gregory. "Evidence-Based Darwinism." *Biological Theory* 5 (2010): 289–91.

Radick, Gregory. "Race and Language in the Darwinian Tradition (and What Darwin's Language–Species Parallels Have to Do with It)." *Studies in History and Philosophy of Biological and Biomedical Sciences* 39 (2008): 359–70.

Radick, Gregory. "[Review of Desmond and Moore, *Darwin's Sacred Cause*]." *Times Higher Education* (2009): 48–49.

Raj, Kapil. "Beyond Postcolonialism . . . and Postpositivism: Circulation and the Global History of Science." *Isis* 104 (2013): 337–47.

Raj, Kapil. "Mapping Knowledge Go-Betweens in Calcutta, 1770–1820." In *The Brokered World: Go-Betweens and Global Intelligence, 1770–1820*, edited by Simon Schaffer, Stefan Collini, Richard Whatmore, and Brian Young, 105–50. Sagamore Beach: Watson, 2009.

Ratnapalan, Laavanyan. "E. B. Tylor and the Problem of Primitive Culture." *History and Anthropology* 19 (2008): 131–42.

Raychaudhuri, Tapan. *Europe Reconsidered: Perceptions of the West in Nineteenth-Century Bengal.* New Delhi: Oxford University Press, 2002.

Reade, W. Winwood. *The Martyrdom of Man.* London: Trübner, 1872.

Revelle, William. "Sir Francis Galton." In *The Encyclopedia of Clinical Psychology*, edited by Robin L. Cautin and Scott O. Lilienfeld, 3:1305–8. 5 vols. Chichester: Wiley Blackwell, 2015.

Richard, Nathalie. "Archeology, Biology, Anthropology: Human Evolution according to Gabriel de Mortillet and John Lubbock." *History and Philosophy of the Life Sciences* 34, nos. 1–2 (2012): 9–31.

Richards, Evelleen. *Darwin and the Making of Sexual Selection.* Chicago: University of Chicago Press, 2017.

Richards, Robert. *Darwin and the Emergence of Evolutionary Theories of Mind and Behavior*. Chicago: University of Chicago Press, 1887.

Richards, Robert. "Darwin on Mind, Morals and Emotions." In *The Cambridge Companion to Darwin*, edited by Jonathan Hodge and Gregory Radick, 96–119. 2nd ed. Cambridge: Cambridge University Press, 2009.

Richards, Robert. "Descent of Man." *American Scientist* 97 (2009): 415–17.

Richardson, Angelique, ed. *After Darwin: Animals, Emotions, and the Mind.* New York: Rodopi, 2013.

Richardson, Angelique. "'The Book of the Season': The Conception and Reception of Darwin's *Expression.*" In *After Darwin: Animals, Emotions, and the Mind*, edited by Angelique Richardson, 51–88. New York: Rodopi, 2013.

Risley, Herbert Hope. *The People of India.* Calcutta: Thacker, Spink, 1915.

Risley, Herbert Hope. "The Study of Ethnology in India." *Journal of the Anthropological Institute of Great Britain and Ireland* 20 (1891): 235–63.

Robertson, William. *The History of America.* 3 vols. Dublin: Messrs Whitestone, 1777–1796.

Rock, David. "The British in Argentina: From Informal Empire to Postcolonialism." *Bulletin of Latin American Research* 24 (2008): 49–77.

Rock, David. "The British in Argentina: From Informal Empire to Postcolonialism." In *Informal Empire in Latin America: Culture, Commerce, and Capital*, edited by Matthew Brown, 49–77. Oxford: Blackwell, 2008.

Rodger, N. A. M. *The Command of the Ocean: A Naval History of Britain, 1649–1815.* London: Penguin Book, 2006.

Romanes, George. *Animal Intelligence.* New York: Appleton, 1891.

Romanes, George. *Mental Evolution in Animals.* New York: Appleton, 1894.

Romanes, George. *Mental Evolution in Man.* New York: Appleton, 1889.

Rosenberg, John D. *Carlyle and the Burden of History.* Oxford: Oxford University Press, 1985.

Rowley-Conwy, Peter. *From Genesis to Prehistory: The Archaeological Three Age System and Its Contested Reception in Denmark, Britain, and Ireland.* Oxford: Oxford University Press, 2007.

Rubiés, Joan-Pau. "New Worlds and Renaissance Ethnology." *History and Anthropology* 6 (1993): 157–97.

Rubiés, Joan-Pau. *Travel and Ethnology in the Renaissance: South India through European Eyes, 1250–1625.* Cambridge: Cambridge University Press, 2000.

Rudwick, Martin. *Bursting the Limits of Time: The Reconstruction of Geohistory in the Age of Revolution.* Chicago: University of Chicago Press, 2005.

Rudwick, Martin. *Earth's Deep History: How It Was Discovered and Why It Matters.* Chicago: University of Chicago Press, 2014.

Rudwick, Martin. *Worlds before Adam: The Reconstruction of Geohistory in the Age of Reform.* Chicago: University of Chicago Press, 2008.

Ruse, Michael. *Monad to Man: The Concept of Progress in Evolutionary Biology.* Cambridge, MA: Harvard fUniversity Press, 1996.

Sadan, Mandy. *Being and Becoming Kachin: Histories beyond the State in the Borderworlds of Burma.* Oxford: Oxford University Press, 2013.

Sarkar, Sumit. *Beyond Nationalist Frames: Postmodernism, Hindu Fundamentalism, History.* Bloomington: Indiana University Press, 2002.

Sarmiento, Domingo F. *Facundo o civilización i barbarie en las pampas argentinas.* Santiago: Imprenta del Progreso, 1845.

Schabas, Margaret. *The Natural Origins of Economics.* Chicago: University of Chicago Press, 2005.

Schaffer, Simon. "The Asiatic Enlightenments of British Astronomy." In *The Brokered World: Go-Betweens and Global Intelligence, 1770–1820*, edited by Simon Schaffer, Lissa Roberts, Kapil Raj, and James Delbourgo, 49–104. Sagamore Beach: Watson, 2009.

Schaffer, Simon. "Newton on the Beach: The Information Order of *Principia Mathematica.*" *History of Science* 47 (2009): 243–76.

Schaffer, Simon, Lissa Roberts, Kapil Raj, and James Delbourgo, eds. *The Brokered World: Go-Betweens and Global Intelligence, 1770–1820.* Sagamore Beach: Watson, 2009.

Scheckner, Peter, ed. *An Anthology of Chartist Poetry.* London: Associated University Presses, 1989.

Schmerling, Philippe Charles. *Recherches Sur Les Ossemens Fossiles Découverts Dans Les Cavernes de La Province de Liège.* Liège: P. J. Collardin, 1833.

Scott, Walter. *Ivanhoe.* Edited by Graham Tulloch. London: Penguin, 2000.

Scott, Walter. *Waverley; or, 'Tis Sixty Years Since.* Edited by Claire Lamont. Oxford: Oxford University Press, 1986.

Secord, James. "Introduction." In *Principles of Geology [Selections from 1830–33]*, by Charles Lyell, ix–xliii. London: Penguin Classics, 1997.

Secord, James. "Knowledge in Transit." *Isis* 95 (2004): 654–72.

Secord, James. *Victorian Sensation: The Extraordinary Publication, Reception, and Secret Authorship of* Vestiges of the Natural History of Creation. Chicago: University of Chicago Press, 2000.

Sen, Keshab Chandra. "Philosophy and Madness in Religion." In *Keshab Chandra Sen and the Brahma Samaj*, edited by T. E. Slater, 84–91. Madras: Society for Promoting Christian Knowledge, 1884.

Sera-Shriar, Efram. "Arctic Observers: Richard King, Monogenism and the Historicisation of Inuit through Travel Narratives." *Studies in History and Philosophy of Biological and Biomedical Sciences* (2015): 23–31.

Sera-Shriar, Efram. "Ethnology in the Metropole: Robert Knox, Robert Gordon Latham and Local Sites of Observational Training." *Studies in History and Philosophy of Biological and Biomedical Sciences* 42 (2011): 486–96.

Sera-Shriar, Efram. "Human History and Deep Time in Nineteenth-Century British Sciences: An Introduction." *Studies in History and Philosophy of Biological and Biomedical Sciences* 51 (2015): 19–22.

Sera-Shriar, Efram. *The Making of British Anthropology, 1813–1871*. London: Pickering and Chatto, 2013.

Sera-Shriar, Efram. "Observing Human Difference: James Hunt, Thomas Huxley, and Competing Disciplinary Strategies in the 1860s." *Annals of Science* 70 (2013): 461–91.

Sera-Shriar, Efram. "Tales from Patagonia: Phillip Parker King and Early Ethnographic Observation in British Ethnology, 1826–1830." *Studies in Travel Writing* 19 (2015): 204–23.

Sera-Shriar, Efram. "What Is Armchair Anthropology? Observational Practices in Nineteenth-Century British Human Sciences." *History of the Human Sciences* 27 (2014): 26–40.

Sevilla Pérez, Elisa. "Imperios informales y naciones poscoloniales: La autoridad de la ciencia." *Historica Critica* (2013): 79–103.

Sheets-Pyenson, Susan. *John William Dawson: Faith, Hope and Science.* Montreal: McGill-Queen's University Press, 1996.

Simms, Brendan. *Three Victories and a Defeat: The Rise and Fall of the First British Empire.* London: Penguin Books, 2008.

Simpson, Thomas. "'Clean out of the Map': Knowing and Doubting Space at India's High Imperial Frontiers." *History of Science* 55, no. 1 (2017): 3–36.

Simpson, Thomas. "Colonial Frontiers in North-Western and North-Eastern British India during the Nineteenth Century." PhD diss., University of Cambridge, 2015.

Singh, C. *Life and Teachings of Swami Dayanand Saraswati.* New Delhi: Jan Gyan Prakashan, 1971.

Sivasundaram, Sujit, ed. "Global Histories of Science." Focus Section. *Isis* 101 (2010): 95–158.

Sivasundaram, Sujit. "Imperial Transgressions: The Animal and Human in the Idea of Race." *Comparative Studies of South Asia, Africa and the Middle East* 35 (2015): 156–72.

Sivasundaram, Sujit. *Nature and the Godly Empire: Science and Evangelical Mission in the Pacific, 1795–1850*. Cambridge: Cambridge University Press, 2005.

Smith, Crosbie. *The Science of Energy: A Cultural History of Energy Physics in Victorian Britain.* Chicago: University of Chicago Press, 1998.

Smith, Roger. "Alfred Russel Wallace: Philosophy of Nature and Man." *British Journal for the History of Science* 6 (1972): 177–99.

Smith, Worthington George. *Man, the Primeval Savage: His Haunts and Relics from the Hill-Tops of Bedfordshire to Blackwall.* London: E. Stanford, 1894.

Sober, Elliott. *Evidence and Evolution: The Logic behind the Science.* Cambridge: Cambridge University Press, 2008.

Sommer, Marianne. *Bones and Ochre: The Curious Afterlife of the Red Lady of Paviland.* Cambridge, MA: Harvard University Press, 2007.

Spearing, H. G. *The Childhood of Art.* New York: G. P. Putnam's Sons, 1912.

Spencer, Herbert. "Progress: Its Law and Cause." *Westminster Review* 67, no. 132 (April 1857): 445–85.

Spencer, Herbert. *Social Statics: Or, the Conditions Essential to Human Happiness Specified, and the First of Them Developed.* London: John Chapman, 1851.

Squier, George E. *Notes on Central America.* New York: Harper and Brothers, 1855.

Stafford, Robert A. *Scientist of Empire: Sir Roderick Murchison, Scientific Exploration and Victorian Imperialism.* Cambridge: Cambridge University Press, 1989.

Stamos, David N. *Darwin and the Nature of Species.* New York: SUNY Press, 2012.

Stauss, David. *Percival Lowell: The Culture and Science of a Boston Brahmin.* Cambridge, MA: Harvard University Press, 2001.

Stepan, Nancy. *The Hour of Eugenics: Race, Gender and Nation in Latin America.* Ithaca: Cornell University Press, 1991.

Stepan, Nancy. *The Idea of Race in Science: Great Britain, 1800–1960*. London: Macmillan, 1982.

Stevenson, John. "Observations on the Marathi Language." *Journal of the Royal Asiatic Society of Great Britain and Ireland* 7 (1843): 84–91.

Stewart, Dugald. *Biographical Memoir of Adam Smith.* New York: Augustus Kelley, 1966.

Stocking, George. *After Tylor: British Social Anthropology, 1888–1951.* Madison: University of Wisconsin Press, 1995.

Stocking, George. "'Cultural Darwinism' and 'Philosophical Idealism' in E. B. Tylor." In *Race, Culture, and Evolution: Essays in the History of Anthropology*, edited by George Stocking, 91–109. Chicago: University of Chicago Press, 1968.

Stocking, George. "Edward Burnett Tylor and the Mission of Primitive Man," In *Delimiting Anthropology: Occasional Essays and Reflections*, edited by George Stocking, 103–15. Madison: University of Wisconsin Press, 2001.

Stocking, George. *Victorian Anthropology*. New York: Free Press, 1987.

Stocking, George. "What's in a Name? The Origins of the Royal Anthropological Institute (1837–71)." *Man* 6 (1971): 369–90.

Strauss, David. *Percival Lowell: The Culture and Science of a Boston Brahmin*. Cambridge, MA: Harvard University Press, 2001.

Streets, Heather. *The Military, Race and Masculinity in British Imperial Culture, 1857–1914*. Manchester: Manchester University Press, 2004.

Strenski, Ivan. *Thinking about Religion: An Historical Introduction to Theories of Religion*. Oxford: Blackwell, 2006.

Strenski, Ivan. *Understanding Theories of Religion: An Introduction*. Chichester: John Wiley and Sons, 2015.

Sulivan, Henry Norton, ed. *Life and Letters of the Late Admiral Sir Bartholomew James Sulivan, K.C.B., 1810–1890*. London: John Murray, 1896.

Stuurman, Siep. "François Bernier and the Invention of Racial Classification." *History Workshop Journal* 50 (Autumn 2000): 1–21.

Tarlach, Gemma. "The Origins of Dogs." *Discover* (December 2016), 32–39.

Taylor, Michael A. *Hugh Miller: Stonemason, Geologist, Writer*. Edinburgh: NMS Enterprises, 2007.

Taylor, Michael W., ed. *Herbert Spencer: Legacies*. Abingdon: Routledge, 2015.

Taylor, Richard. *The Past and Present of New Zealand with Its Prospects for the Future*. London: William Macintosh, 1868.

Temple, Frederick. *The Relations between Religion and Science: Eight Lectures Preached before the University of Oxford in the Year 1884*. New York: Macmillan, 1884.

Tilley, Helen. "Global Histories, Vernacular Science, and African Genealogies; or, Is the History of Science Ready for the World?" *Isis* 101 (2010): 110–19.

Trautmann, Thomas. *Aryans and British India*. Berkeley: University of California Press, 1997.

Thompson, Andrew. *The Empire Strikes Back? The Impact of Imperialism on Britain from the Mid-Nineteenth Century*. London: Pearson-Longman, 2005.

Topham, Jonathan. "Beyond the 'Common Context': The Production and Reading of the Bridgewater Treatises." *Isis* 89, no. 2 (1998): 233–62.

Topham, Jonathan. "Science and Popular Education in the 1830s: The Role of the 'Bridgewater Treatises.'" *British Journal for the History of Science* 25, no. 4 (1992): 397–430.

Topham, Jonathan. "Scientific Publishing and the Reading of Science in Nineteenth-Century Britain: A Historiographical Survey and Guide to the Sources." *Studies in the History and Philosophy of Science* 31 (2000): 559–612.

Trevor-Roper, Hugh. *The Last Days of Hitler*. London: Macmillan, 1947.

Turner, Frank. *Between Science and Religion: The Reaction to Scientific Naturalism in Late Victorian England.* New Haven: Yale University Press, 1974.

Turner, Frank. "The Victorian Conflict between Science and Religion: A Professional Dimension." *Isis* 69 (1978): 356–76.

Turner, George. *Nineteen Years in Polynesia: Missionary Life, Travel and Researches in the Islands of the Pacific.* London: John Snow, 1861.

Tylor, Edward Burnett. *Anahuac: or, Mexico and the Mexicans, Ancient and Modern.* London: Longman, Green, Longman, and Roberts, 1861.

Tylor, Edward Burnett. *Anthropology: An Introduction to the Study of Man and Civilisation.* London: Macmillan, 1881.

Tylor, Edward Burnett. *Primitive Culture: Researches into the Development of Mythology, Philosophy, Religion, Language, Art and Custom.* 2 vols. London: John Murray, 1871.

Tylor, Edward Burnett. *Primitive Culture: Researches into the Development of Mythology, Philosophy, Religion, Language, Art and Custom.* 2nd ed. 2 vols. London: Henry Holt, 1889.

Tylor, Edward Burnett. *Researches into the Early History of Mankind and the Development of Civilisation.* London: John Murray, 1865.

Van Riper, A. B. *Men among the Mammoths: Victorian Science and the Discovery of Human Prehistory*. Chicago: University of Chicago Press, 1993.

van Wyhe, John. *Dispelling the Darkness: Voyage in the Malay Archipelago and the Discovery of Evolution by Wallace and Darwin.* Singapore: World Scientific Publishing, 2013.

van Wyhe, John, and Peter Kjærgaard. "Going the Whole Orang: Darwin, Wallace and the Natural History of Orangutans." *Studies in History and Philosophy of Biology and Biomedical Sciences* 51 (2015): 53–63.

Vetter, Jeremy. "The Unmaking of an Anthropologist: Wallace Returns from the Field, 1862–70." *Notes and Records of the Royal Society* 64 (2009): 25–42.

Vetter, Jeremy. "Wallace's Other Line: Human Biogeography and Field Practice in the Eastern Colonial Tropics." *Journal of the History of Biology* 39 (2006): 89–123.

Villavicencio, Manuel. *Geografía de La Republica de Ecuador.* New York: Robert Graighead, 1858.

Vivekananda, Swami. *The Complete Works.* 8 vols. Calcutta: Advaita Ashrama, 1973.

Wallace, Alfred Russel. *Contributions to the Theory of Natural Selection: A Series of Essays.* London: Macmillan, 1870.

Wallace, Alfred Russel. *Darwinism.* London: Macmillan, 1889.

Wallace, Alfred Russel. "Government Aid to Science." *Nature* 1 (1870): 288–89.

Wallace, Alfred Russel. *My Life: A Record of Events and Opinions.* 2 vols. London: Chapman and Hall, 1905.

Wallace, Alfred Russel. *Narrative of Travels on the Amazon and Rio Negro, with an Account of the Native Tribes, and Observations on the Climate, Geology, and Natural History of the Amazon Valley.* London: Reeve, 1853.

Wallace, Alfred Russel. "Notes of a Journey Up the Sadong River, in North-west Borneo." *Proceedings of the Royal Geographical Society of London* 6 (1856): 193–205.

Wallace, Alfred Russel. "On the Law which has Regulated the Introduction of New Species." *Annals and Magazine of Natural History, including Zoology, Botany, and Geology* 16 (1855): 184–96.

Wallace, Alfred Russel. "On the Natural History of the Aru Islands." *Annals and Magazine of Natural History* 20 (1857): 473–85.

Wallace, Alfred Russel. *On the Organic Law of Change. A Facsimile Edition and Annotated Transcription of Alfred Russel Wallace's Species Notebook of 1855–1859.* Edited by James Costa. Cambridge, MA: Harvard University Press, 2013.

Wallace, Alfred Russel. "On the Tendency of Varieties to Depart Indefinitely from the Original Type." *Proceedings of the Linnean Society of London* 3, no. 9 (1858): 53–62.

Wallace, Alfred Russel. "On the Varieties of Man in the Malay Archipelago." *Transactions of the Ethnological Society of London* 3 (1865): 196–215.

Wallace, Alfred Russel. "The Origin of Human Races and the Antiquity of Man Deduced from the Theory of 'Natural Selection.'" *Journal of the Anthropological Society of London* 2 (1864): clviii–clxxxvii.

Wallace, Alfred Russel. "Primitive Culture: Researches into the Development of Mythology, Philosophy, Religion, Art, and Custom." *The Academy* 3 (1872): 69–71.

Wallace, Alfred Russel. *The Scientific Aspect of the Supernatural.* London: F. Farrah, 1866.

[Wallace, Alfred Russel]. "Sir Charles Lyell on Geological Climates and the Origin of Species." *Quarterly Review* 126, no. 252 (1869): 359–94.

Walpole, Spencer. *A History of England from the Conclusion of the Great War in 1815*. 6 vols. London: Longmans, Green, 1890.

Wartelle, Jean-Claude. "La Société d'Anthropologie de Paris de 1859 à 1920." *Revue d'Histoire des Sciences Humaines* 10, no. 1 (2004): 125–71.

Wheeler, Roxann. *The Complexion of Race: Categories of Difference in Eighteenth-Century British Culture*. Philadelphia: University of Pennsylvania, 2000.

Wheeler-Barclay, Marjorie. *The Science of Religion in Britain, 1860–1915*. Charlottesville: University of Virginia Press, 2010.

White, Andrew Dickson. *A History of the Warfare of Science with Theology in Christendom*. 2 vols. New York: D. Appleton, 1896.

White, Paul. "Darwin's Emotions: The Scientific Self and the Sentiment of Objectivity." *Isis* 100 (2009): 811–26.

White, Paul. *Thomas Huxley: Making the "Man of Science."* Cambridge: Cambridge University Press, 2003.

Wills, Garry. "Lincoln's Black History." *New York Review of Books* (11 June 2009): 52–55.

Wilson, Leonard G. *Sir Charles Lyell's Scientific Journals on the Species Question*. New Haven: Yale University Press, 1970.

Winterbottom, Anna. "Medicine and Botany in the Making of Madras, 1680–1720." In *The East India Company and the Natural World*, edited by Vinita Damodaran, Anna Winterbottom, and Alan Lester, 35–57. Basingstoke: Palgrave Macmillan, 2014.

Worster, Donald. *Nature's Economy: A History of Ecological Ideas*. Cambridge: Cambridge University Press, 1994.

Zastoupil, Lynn. *Rammohun Roy and the Making of Victorian Britain*. Basingstoke: Palgrave Macmillan, 2010.

Zeller, Suzanne. "Environment, Culture, and the Reception of Darwin in Canada, 1859–1909." In *Disseminating Darwinism: The Role of Place, Race, Religion, and Gender*, edited by Ronald L. Numbers and John Stenhouse, 91–122. Cambridge: Cambridge University Press, 1999.

Zeller, Suzanne. *Inventing Canada: Early Victorian Science and the Idea of a Transcontinental Nation*. Carleton Library. Montreal: McGill-Queen's University Press, 2009.

LIST OF CONTRIBUTORS

ABIGAIL NIEVES DELGADO works as a researcher at the Centre for Anthropological Knowledge of Scientific and Technological Cultures (CAST) at Ruhr University, Bochum, Germany. She received her PhD from the National Autonomous University of Mexico (UNAM) in 2016. Her research focuses on the history of physical anthropology and on current practices in forensic sciences and biometric identification. Her current project, which is funded by the German Academic Exchange Service (DAAD), deals with the use of racial categories in facial recognition technologies.

MAURIZIO ESPOSITO is head of the Department of Philosophy at the University of Santiago, Chile. He has previously worked at UNAM, Mexico, after completing his PhD at the University of Leeds, UK, in 2011. He studied history and philosophy of science at the University of Bologna, Italy, at the University of Leuven, Belgium, and at the University of Montreal, Canada. He is primarily interested in the history and philosophy of the life sciences during the nineteenth and twentieth centuries. He has published a monograph and articles on the history of biology and on the relationship between science and society. He is also interested in the history of science in Latin America and the relationship between imperialism and scientific exploration in the nineteenth century, particularly in South America.

IAN HESKETH is an Australian Research Council Future Fellow in the Institute for Advanced Studies in the Humanities at the University of Queensland, Australia. He works on the relationship between science, religion, and history. His books include *Of Apes and Ancestors: Evolution, Christianity, and the Oxford Debate* (2009), *The Science of History in Victorian Britain* (2011), and most recently, *Victorian Jesus: J. R. Seeley, Religion, and the Cultural Significance of Anonymity* (2017).

NANNA KATRINE LÜDERS KAALUND received her PhD in science and technology studies at York University, Canada, in 2017. She is currently a lecturer in the history of science communication at the Centre for History and Philosophy of Science at the University of Leeds and currently holds a visiting research fellowship at the Leeds Centre for Victorian Studies at Leeds Trinity University. Kaalund has published on print history and the historical relationship between science and religion and is currently working toward turning her doctoral dissertation, "From Science in the Arctic to Arctic Science: A Transnational Study of Arctic Travel Narratives, 1818–1883," into a monograph.

HELEN KINGSTONE is currently a lecturer in English literature (Victorian Studies) at the University of Glasgow. Her research addresses the relationship between memory and history in the nineteenth century, focusing on how different writers in genres and forms approached contemporary history. Her monograph, *Victorian Narratives of the Recent Past: Memory, History, Fiction*, was published by Palgrave Macmillan in 2017. She has also published articles on the conflicting demands of feminism and nationalism for the Irish historian Alice Stopford Green (*Journal of Victorian Culture* 19, no. 4), and on the persistence of religious motifs in fin-de-siècle utopian fiction.

THEODORE KODITSCHEK is professor emeritus of history at the University of Missouri. He is the author of numerous articles and several books, most recently, *Liberalism, Imperialism and the Historical Imagination: Nineteenth Century Visions of a Greater Britain* (Cambridge University Press, 2011). He is currently working on a book to be tentatively titled *Race and Liberal Imperial Capitalism: Greater Britain, 1030–1914*.

CHRIS MANIAS is lecturer in the history of science and technology at King's College London. He studied for his PhD at Birkbeck, University of London (awarded 2008) and has previously held academic positions at the German Historical Institute London and the Universities of Bristol, Exeter, and Manchester. His book, *Race, Science and the Nation: Reconstructing the Ancient Past in Britain, France and Germany, 1800–1914*, was published in 2013 by Routledge. He is currently working on a new project titled *The Lost Beasts: International Palaeontology and the Evolution of the Mammals, 1880–1950*.

GREGORY RADICK is professor of history and philosophy of science at the University of Leeds. His books include *The Simian Tongue: The Long Debate about Animal Language* (Chicago, 2007), which was awarded the Suzanne J. Levinson Prize of the History of Science Society; and as co-editor with Jonathan Hodge, *The Cambridge Companion to Darwin* (Cambridge, 2003; 2nd ed., 2009). He is working on a new book about the circa 1900 debate over Mendelism. He is past president of the British Society for the History of Science and currently president-elect of the International Society for the History, Philosophy and Social Studies of Biology.

EFRAM SERA-SHRIAR is lecturer in modern history at Leeds Trinity University. His research explores the intersection of voyages of exploration, race, visual culture, science, medicine, and society throughout the British Empire from the late eighteenth century to the early twentieth century. He has published extensively on the history of the human sciences, including his monograph, *The Making of British Anthropology, 1813–1871* (2013). Since 2006 he has been an editor on the John Tyndall Correspondence Project. In his current research, he is examining anthropology's engagement with the modern spiritualist movement during the late Victorian era.

THOMAS SIMPSON is junior research fellow in history at Gonville and Caius College, University of Cambridge. He is currently completing a monograph on the concept of the frontier in nineteenth-century colonial India. His research spans imperial history, the history of science, and historical geography, and his publications include articles in the *Historical Journal* and *History of Science* on British border making in India, and surveying and mapping in the age of high empire.

INDEX